THE LIVES OF
FROGS

THE LIVES OF FROGS
A NATURAL HISTORY

Jim Labisko &
Richard Griffiths

PRINCETON UNIVERSITY PRESS
PRINCETON AND OXFORD

Published by Princeton University Press
41 William Street, Princeton, New Jersey 08540
99 Banbury Road, Oxford OX2 6JX
press.princeton.edu

Copyright © 2025 by UniPress Books Limited
www.unipressbooks.com

All rights reserved. No part of this book may be reproduced or transmitted in any form or by any means, electronic or mechanical, including photocopying, recording, or by any information storage-and-retrieval system, without written permission from the copyright holder. Requests for permission to reproduce material from this work should be sent to permissions@press.princeton.edu

Library of Congress Control Number 2024944461
ISBN 978-0-691-25501-9
Ebook ISBN 978-0-691-25506-4

Typeset in Bembo and Futura

Printed and bound in Malaysia
10 9 8 7 6 5 4 3 2 1

British Library Cataloging-in-Publication Data is available

This book was conceived, designed, and produced by
UniPress Books Limited
Publisher: Jason Hook
Commissioning editor: Kate Shanahan
Project manager: Richard Webb
Designer & art director: Wayne Blades
Picture researcher: Elaine Willis
Illustrator: John Woodcock
Maps: Les Hunt

Cover images: (Front): Safiqul BD / Shutterstock;
(spine and back): Nynke van Holten / Shutterstock

CONTENTS

6
INTRODUCTION

18
THE MAKING OF MODERN SURVIVORS

42
LIFE ON LAND & IN WATER

74
COMMUNICATION & REPRODUCTION

106
EGGS, TADPOLES, & PARENTING

136
GETTING AROUND

166
FROGS, FOOD, & FEEDING

190
DEFENSES AGAINST PREDATORS & PATHOGENS

220
THE UPS & DOWNS OF FROG POPULATIONS

248
FROGS IN THE FUTURE

280 Glossary
282 Resources
284 Index
288 Acknowledgments

INTRODUCTION

The world of frogs

The Vallée de Mai Nature Reserve on the Seychelles island of Praslin is like nowhere else in the world. A surviving remnant of ancient palm forest, indicating what these forests may have looked like millions of years ago, this habitat is one of the most important on the planet and a designated UNESCO World Heritage Site. Unlike many other islands where the native trees have been supplanted by introduced invasives, the Vallée de Mai supports no fewer than six Seychelles' endemic palm species. The most spectacular of these is the coco-de-mer (*Lodoicea maldivica*) which produces the largest seed of any plant in the world. Indeed, almost exactly the size and shape of a human bottom, coco-de-mer seedpods liven up the décor of bars and hotels throughout the islands. Circling the pillar-like trunks of the palm trees are emerald-green *Phelsuma* day geckos, while among the foliage Seychelles Black Parrots (*Coracopsis barklyi*) whistle and squabble noisily with each other. But we were not at the Vallée de Mai to admire the breathtaking landscape or its unique reptile and bird fauna. We were there to search for an animal that is small, behaviorally cryptic, and (arguably) rather unspectacular in appearance. Indeed, to the casual observer, the Seychelles Frog (*Sooglossus sechellensis*, see page 40) might look like any other small, brown frog. For us, however, this unassuming animal was the holy grail—a piece of the puzzle that might provide the answer to understanding long-argued relationships within the frog tree of life. In fact, few have ever seen this frog, and its calls were thought to be from a "whistling snake." If we could find this frog, record its unique vocalization, and capture some DNA, we might discover not only how it is related to other frogs in the Seychelles, but also something about how ancient faunas came to colonize this remote part of the world.

→ The Bornean Rainbow Toad (*Ansonia latidisca*) was rediscovered in 2011, having not been reported since the 1920s.

↓ Previously found only on the islands of Mahé and Silhouette, the Seychelles Frog (*Sooglossus sechellensis*) has been known to science since the 1890s. In 2009 it was discovered on a third island, Praslin.

INTRODUCTION

By observing the habitats in which we found the frogs, we could potentially shed light on how they have coped with historic climate change and how frogs in general might fare in the future. But these analyses were for the months ahead back in the lab and head down at the desk. Here and now, it was more about sifting through piles of fallen palm leaves, getting muddy in damp hollows, straining our ears for frog calls, and groveling on our bellies among massive granite boulder fields. These same basic methods are used to search for frogs across the world, and we are not aware of any scientific advances that have yet to replace such fieldcraft.

You don't have to travel to exotic forests to find frogs, however. Indeed, as we aim to show in the pages of this book, frogs can be found all over the world and in many different places: from the bottom of lakes to the tops of trees, from deserts to the Arctic tundra, and from urban parks and gardens to the world's

↑ The Madagascar Bright-eyed Frog (*Boophis madagascariensis*) is a distinctive frog with dermal flaps on its elbows and heels. These likely enhance the frogs' camouflage by diffusing the outline of its body.

→ Sylvia's Tree Frog (*Cruziohyla sylviae*) from Central America is well camouflaged when at rest in the rainforest canopy, but displays its striking orange and black-barred coloration when active.

INTRODUCTION

great wildernesses. All of these places have their own distinctive and perfectly adapted frogs, some of them small and well camouflaged like our quarry in the Seychelles and others spectacularly colored in shades of red, green, blue, and yellow. Some frogs are regarded as delicacies or used for medicinal purposes, while others contain chemicals that are among the most toxic known to humans.

The life histories of frogs are as diverse as their form, function, and distribution, and some are outright bizarre. What other animal group has such a unique combination of species that can hop, leap, swim, or glide; that swallow their young in order to raise them; has females that entrust their eggs to a male to look after; and aggressive males that will attack humans if they come near their tadpoles? The following chapters offer a global tour of the wonderful diversity of frogs.

Frogs are the great survivors, with ancestors that saw the rise and fall of the dinosaurs and great climatic upheaval, but today they are in deep trouble. Many species have already succumbed to threats such as habitat loss and emerging diseases, and it is estimated that 41 percent of all amphibians are threatened with extinction—a higher figure than mammals and birds, which have been much better studied. As scientists, we spend a lot of time unravelling the important role that frogs play in wider ecosystems and the "value" they have to society. However, we believe the most important reason for conserving frogs is simply because they are wonderful, inspiring, and fascinating animals—things you cannot put a price on. This book is intended to be a celebration of the lives and diversity of frogs. We hope you will appreciate that frogs are important and worth protecting just because they are frogs. And, yes, we did find the elusive frogs from Vallée de Mai.

Jim Labisko & Richard Griffiths

← The Arroyo Toad (*Anaxyrus californicus*) is an endangered desert species endemic to southwestern California and the Baja California region of Mexico.

INTRODUCTION

What are frogs and toads?

Frogs and toads belong to the order Anura and, strictly speaking, all toads are frogs. In general, frogs are lithe, long-legged, jumpy amphibians with a smooth, damp skin, while toads are squat, have a dry, warty appearance, and hop or walk on shorter legs. That said, these features have evolved independently several times, meaning that some families contain closely related species of both frogs and toads.

→ Typical "true" toads (such as this American Toad, *Anaxyrus americanus*) are characterized by dry, warty skins and relatively short back legs that mean they walk and hop rather than leap.

↓ Typical "true" frogs such as the European Common Frog, *Rana temporaria*) have smooth, moist skins and long back legs for jumping.

The designation of an amphibian as a frog or a toad probably originates in Europe, where the two anurans have historically been separated according to their observable features. The Anglo-Saxon terms *frogga* or *froskr*, and *tade* or *tadde*, may be precursors of the present-day "frog" and "toad," respectively. Indeed, when "tade" is combined with "poll"—an ancient word for "head"—we can derive "tadpole," which is an apt description for the larval stage of

WHAT ARE FROGS AND TOADS?

frogs and toads, looking as they do like a large, oval head propelled by a tail. An alternative but little-used name for a tadpole is "pollywog" (sometimes spelled "polliwog" or "pollywig"). This may stem from the Middle English term *polwygle*, meaning "head that wiggles."

The Latin terms for frog and toad, "rana" and "bufo," are still widely used, and reflected in the names of two families that are sometimes referred to as the true frogs, Ranidae (see the European Common Frog, *Rana temporaria*, page 32), and the true toads, Bufonidae (see the American Toad, *Anaxyrus americanus*, page 34). The word "bufo" is probably derived from a German verb meaning "to puff," which could relate to the ability of some toads to inflate their bodies when threatened by a predator.

Given there is little scientific justification for distinguishing between frogs and toads, why do biologists sometimes recognize them as quite different animals? The answer probably lies in the fact that many frogs are smooth-skinned (and so potentially a source of food for humans) whereas the wartier toads are not. Indeed, the "warts" of toad skin contain toxins that can be highly detrimental to a predator. That said, there are also many highly toxic, non-warty frogs.

Only one family—the Bufonidae, or true toads—contains species exclusively regarded as toads. What can be even more confusing is that the terms "frog" or "toad" are sometimes used interchangeably for the same species, the African Clawed Frog or Clawed Toad (*Xenopus laevis*) being a good example, although it is aquatic and has a smooth, slippery skin. Members of another family, Megophryidae (see the Long-nosed Horned Frog, *Pelobatrachus nasutus*, page 214) are sometimes even referred to as toad frogs.

INTRODUCTION

Counting and classifying frogs

At the time of writing, there are 8,715 amphibian species, of which 7,678 are frogs. Salamanders currently number 816 and caecilians (a clade of highly specialized, limbless amphibians) 221. The number of frog species is actually increasing as scientists explore remote areas of the world and use an array of novel techniques to unravel their relationships.

Taxonomy assesses the characteristics (traits) of an organism or group of organisms in order to classify them according to shared features, and thus form immediately more inclusive hierarchical levels of taxonomic rank. For example, a higher-level rank for the American Toad is class (Amphibia), followed sequentially by the lower ranks of order (Anura), family (Bufonidae), genus (*Anaxyrus*), and species (*Anaxyrus americanus*).

Examples of key characteristics used by frog taxonomists include morphology, genetics, vocalizations, ecology, and geographic distributions. We currently have a good understanding of the relationships between the main groups, such as the higher taxonomic ranks of suborder and superfamily. It is the picture in the lower groupings—for example, the relationships within frog families—that is more difficult to resolve. The reasons for this are many, but include rapid radiations in widely distributed, species-rich frog families (for instance, Microhylidae, the narrow-mouthed frogs). This makes it challenging to ascertain relationships within groups and difficult to distinguish similar, often closely related, species that arise due to similar ecological drivers (such as multiple genera of hylid treefrogs). That said, integrative approaches to anuran taxonomy—those that combine multiple methods to understand the evolutionary relationships of anurans—are bearing fruit.

↑ The majority of frogs, like this Agile Frog (*Rana dalmatina*), consume invertebrate prey such as insects and earthworms.

← The Bush Frog (*Philautus nepenthophilus*; left) is active year-round in the rainforests of Borneo, and uses pitcher plants both as a refuge, and a place to breed, whereas Australia's Holy Cross Frog (*Notaden bennettii*; right) thrives in dry habitat by remaining underground for long periods and is only seen after heavy rains.

Such methods regularly employ assessments of morphology, DNA, and vocalizations, with new technologies such as 3D scans of anatomy and whole-genome sequencing now revealing previously unexplored facets of anuran biology.

CLASSIFYING MODERN FROGS

Modern frogs can initially be divided into two suborders: Archaeobatrachia and Neobatrachia. Archaeobatrachia ("archaeo" from the Greek, meaning "ancient") comprises the oldest known extant anuran lineages and has the longest evolutionary history. Neobatrachia ("neo" from the Greek, meaning "new") contains some of the most speciose groups of anurans known and represents more than 95 percent of extant species.

Archaeobatrachian frogs can be classified as falling into four superfamilies: Leiopelmatoidea, Discoglossoidea,

INTRODUCTION

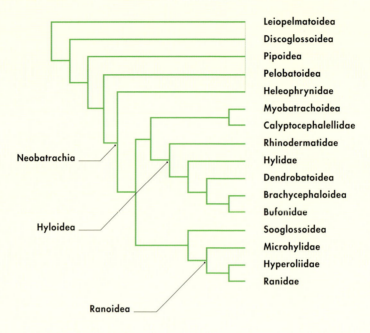

Evolutionary relationships
A cladogram showing the evolutionary relationships of some of the major groups of frogs. Those with the longest evolutionary histories (Leiopelmatoidea, Discoglossoidea, Pipoidea, and Pelobatoidea) are collectively known as Archaeobatrachia, while the most recent frogs form a clade (a group containing the common ancestor and all its descendants) known as Neobatrachia. Names at the tips represent both superfamily and family level clades.

Pipoidea, and Pelobatoidea. Most contain relatively few species, and although pipoid frogs reach well into double figures (41 species), pelobatoids are in the hundreds, primarily due to the ongoing discovery and description of frogs belonging to the Megophryidae (314 species and counting), a widely distributed and diverse family found across South and Southeast Asia.

In Neobatrachia, the majority of frog taxa are grouped within two superfamilies: Hyloidea (around 54 percent of extant frog species) and Ranoidea (around 39 percent of extant frog species). There are several other clades that are closely related to Hyloidea (Calyptocephalellidae, Myobatrachidoidea) and Ranoidea (Sooglossoidea), with one, Heleophrynidae—the oldest known neobatrachian lineage—last sharing a common ancestor with all other neobatrachians around 146 million years ago (Mya). The number of recorded frog species increases year-on-year—in the first six months of 2023 alone, 60 new species were described—with discoveries often due to reappraisal of previously described taxa using new techniques. Knowing how many species there are, and understanding how they are related, plays a vital role in elucidating the evolutionary history of frogs, but it is only part of the story.

A SELECTION OF THE MAJOR CLADES OF FROGS

Here we show a nonexhaustive selection of anuran taxonomic groups, and several of the major clades that comprise them (corresponding with the cladogram on page 16). The number of species within a given taxon is shown in brackets.

SUBORDER: ARCHAEOBATRACHIA (OLD FROGS)

Four superfamilies:

LEIOPELMATOIDEA
ASCAPHIDAE—tailed frogs (2 spp.)

LEIOPELMATIDAE—New Zealand frogs (3 spp.) ↓

DISCOGLOSSOIDEA
ALYTIDAE—midwife toads; painted toads (12 spp.) ↓

PIPOIDEA
RHINOPHRYNIDAE—Mexican burrowing toads (1 sp) ↓

PIPIDAE—tongueless frogs (41 spp.)

PELOBATOIDEA
SCAPHIOPODIDAE—North American spadefoot toads (7 spp.) ↓

PELODYTIDAE—parsley frogs (5 spp.)

PELOBATIDAE—spadefoot toads (6 spp.)

MEGOPHRYIDAE—Asian horned frogs (316 spp.) ↓

SUBORDER: NEOBATRACHIA (NEW FROGS)

Three superfamilies, three families:

HELEOPHRYNIDAE—ghost frogs (6 sp)

MYOBATRACHIDAE—Australian water frogs (136 spp.) ↓

CALYPTOCEPHALELLIDAE—helmeted water toads (5 spp.)

HYLOIDEA—two superfamilial-level clades and 19 families, including:

RHINODERMATIDAE—mouth-brooding frogs (3 spp.)

HYLIDAE—"true" treefrogs; leaf frogs (1,058 spp.) ↓

DENDROBATOIDEA—poison frogs; cryptic poison frogs (346 spp.) ↓

BRACHYCEPHALOIDEA—saddle-back toads; robber frogs; rain frogs; landfrogs (1,254 spp.) ↓

BUFONIDAE—"true" toads (654 spp.) ↓

SOOGLOSSOIDEA—purple frogs; Seychelles frogs (6 spp.)

RANOIDEA—one superfamilial-level clade, and 21 families including:

MICROHYLIDAE—narrow-mouthed frogs (745 spp.) ↓

HYPEROLIIDAE—African reed frogs; running frogs (236 spp.) ↓

RANIDAE—"true" frogs (460 spp.)

THE MAKING OF MODERN SURVIVORS

THE MAKING OF MODERN SURVIVORS

Ancestors: From fish to modern frogs

Something quite astounding happened 375 million years ago, an event of such significance that its evolutionary repercussions have shaped the world we know today. It took place in the Devonian Period (around 419–359 mya), when the climate was warm and global biodiversity was booming. The terrestrial habitat was populated by invertebrates, fungi, and early plants such as ferns, horsetails (*Equisetum*), and giant, tree-sized mosses. This is where the story of frogs, and indeed all tetrapods, begins.

↘ Frogs have been hopping, jumping, walking, swimming, and (more recently) gliding on our planet at least 186 million years longer than humans, which arrived on the scene 2.8 million years ago.

Had you been exploring freshwater biodiversity in the rich, shallow, marshland habitats of the Late Devonian—around 383–359 mya—you may have surmised that an explosion of fish diversity was taking place. Intrigued, you might then have observed that one group in particular, those in the clade Sarcopterygii (lobe-finned fishes), were both looking and behaving a little differently. At this time, sarcopterygian fish were in the process of developing adaptations in the limbs, skull, and teeth which would transform vertebrate mobility and locomotion, paving the way for advances in breathing, feeding, and hearing that would ultimately allow these pioneers to emerge from the swamps and populate the land.

518 MYA CAMBRIAN: First known vertebrate (*Myllokunmingia*)

439 MYA SILURIAN: First cartilaginous fish (Chondrichthyes)

425 MYA SILURIAN: Appearance of bony fish (Osteichthyes)

421 MYA SILURIAN: Lobe-finned bony fish (Sarcopterygii) lineage diverges

370.8 MYA DEVONIAN: Appearance of *Tiktaalik roseae*, the sarcopterygian tetrapodomorph

330 MYA CARBONIFEROUS: Appearance of first amphibians (Temnospondyli)

310 MYA CARBONIFEROUS: Earliest known amniote, and the oldest known reptile (*Hylonomus lyelli*)

250 MYA TRIASSIC: Earliest known lissamphibian (*Triadobatrachus massinoti*)

443.8 MYA: Ordovician-Silurian mass extinction (85% loss)   358.9 MYA: Devonian mass extinction (75% loss)

PALEOZOIC
(538.8–251.9 MYA)

Timeline not to scale

ANCESTORS: FROM FISH TO MODERN FROGS

Tiktaalik roseae represents a major step in the evolution of tetrapods: the transition from water to land.

During your Devonian explorations, if you happened to be present on what is today Ellesmere Island in Nunavut, Canada, you may also have seen a very special animal indeed. The sarcopterygian fish *Tiktaalik roseae*, with its body scales and ray-like fins, jaw, and palate, combined with a shortened skull roof, modified ear, mobile neck, and functional wrist, represents the vertebrate transition from fish-like ancestors to the modern tetrapod (four-footed) body plan. These adaptations signal the vertebrate transition from water to land and the appearance of an extraordinarily diverse range of limbed terrestrial vertebrates—the first amphibians—during the Carboniferous Period (around 359–299 mya), which is also known as the Age of Amphibians. One group, the Temnospondyli, is of particular significance.

Ranging from a few centimeters to more than 20 ft (6 m) in body length, the fossil remains of temnospondyls have been found on every continent. Much like today's amphibians, some were fully aquatic and others terrestrial. However, they differed in several ways from modern amphibians, having scales and extensive bony armor—osteoderms (bony plates in the skin)—most likely to prevent desiccation and to serve as protection against predators. At least one group, the fully aquatic and crocodile-like trematosaurs, are thought to have regularly traversed marine environments.

| 230 MYA TRIASSIC: Earliest known salamander (*Triassurus sixtelae*) | 220 MYA TRIASSIC: Earliest known caecilian (*Funcusvermis gilmorei*) | 205 MYA TRIASSIC: Earliest known mammal (*Morganucodon watsoni*) | 189 MYA JURASSIC: Earliest known frog (*Prosalirus bitis*) | 142 MYA CRETACEOUS: Ancestral lineage of neobatrachian frogs diverges, leading to today's major groups | 123 MYA CRETACEOUS: Earliest known modern frog (*Callobatrachus sanyanensis*) | 66 MYA CRETACEOUS: Explosion of diversity in hyloid and ranoid frogs and the emergence of arboreality |

251.9: MYA Permian-Triassic mass extinction (The Great Dying: 95% loss)   201.4 MYA: Triassic-Jurassic mass extinction (80% loss)   66 MYA: Cretaceous-Paleogene Extinction (78% loss)

MESOZOIC	CENOZOIC
(251.9–66.0 MYA)	(66.0–0.0 MYA)

THE MAKING OF MODERN SURVIVORS

MODERN AMPHIBIAN ANCESTORS

Temnospondyli are the putative ancestors from which the three orders of modern amphibians, Anura: frogs (and toads), Caudata: salamanders (and newts), and Gymnophiona: caecilians—which are collectively known as Lissamphibia—are descended. During the Early Triassic (around 252–247 mya), a remarkably frog-like amphibian was living in what is now Madagascar. Approximately 4 in (10 cm) long, it had a short, salamander-like tail, but several other skeletal characteristics, including one of immediate significance for frogs, singled it out. The hind limbs of this animal were longer, only slightly, but longer nonetheless than its forelimbs—this proto-frog could jump … at least a little! Following its initial identification in 1936, *Triadobatrachus massinoti* was redescribed in 1989, since when its significance as a transitional fossil, the earliest known lissamphibian, and the oldest known member of the lineage that leads directly to modern frogs has been recognized.

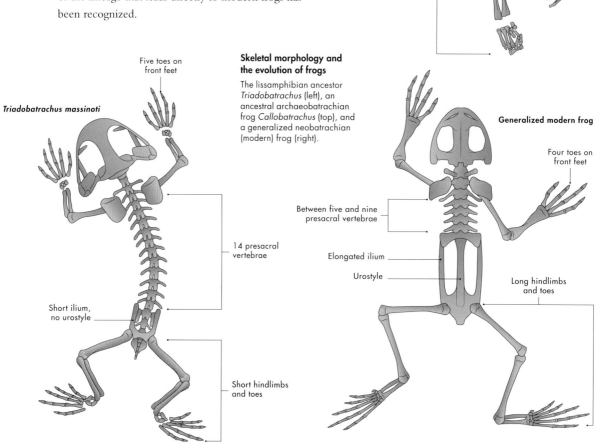

Skeletal morphology and the evolution of frogs

The lissamphibian ancestor *Triadobatrachus* (left), an ancestral archaeobatrachian frog *Callobatrachus* (top), and a generalized neobatrachian (modern) frog (right).

ANCESTORS: FROM FISH TO MODERN FROGS

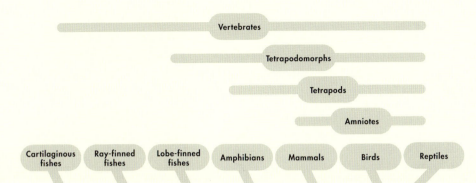

The evolutionary relationships of vertebrates
All vertebrate lineages belong to several inclusive groups, but each group does not include all vertebrates. For example, only amphibians, mammals, birds, and reptiles are tetrapods.

↓ Dorsal view of the only known *Triadobatrachus* fossil, from the Triassic of Madagascar.

Much like *Tiktaalik*, *Triadobatrachus*, with its combination of salamander- and frog-like traits, illuminated a transitional stage in the evolution of amphibians, and represents a link with modern frogs.

The final stop, at least in the deep-time evolutionary history of frogs, lands us well into the Early Cretaceous (around 145–100 mya), in what is today the Yixian Formation in Liaoning, China, and the discovery of one of the oldest known direct ancestors of living frogs to date: the Sanyan Frog (*Callobatrachus sanyanensis*). Described in 1999, its nearly complete remains provided the opportunity for extensive assessment and comparison of its skeletal morphology with that of modern frogs. As a result, *Callobatrachus* is the earliest known ancestor of a family of living anurans, Alytidae.

THE MAKING OF MODERN SURVIVORS

Modern frog distribution

From the Permian to the Jurassic all the continents we know today formed a single landmass: the supercontinent known as Pangaea. Although Pangaea began to fragment around 200 mya, by that time the lineage that gave rise to modern frogs was already well-established. Consequently, several anuran families, including the "true" frogs (Ranidae) and "true" toads (Bufonidae), have continent-spanning distributions, whereas others such as neotropical glass frogs (Centrolenidae) and Australasian water frogs (Myobatrachidae) have more restricted distributions. However, drivers of frog biogeography are diverse and comprise both global and regional events.

→ The wide-ranging European Green Toad (*Bufotes viridis*) is remarkably tolerant of dry conditions. Many bufonids are adept at dispersal, and green toads may travel 3.1 miles (5 km) to reach their annual breeding sites.

→→ Glass frogs are so named because of their translucent ventral skin, through which it's often possible to see their internal organs.

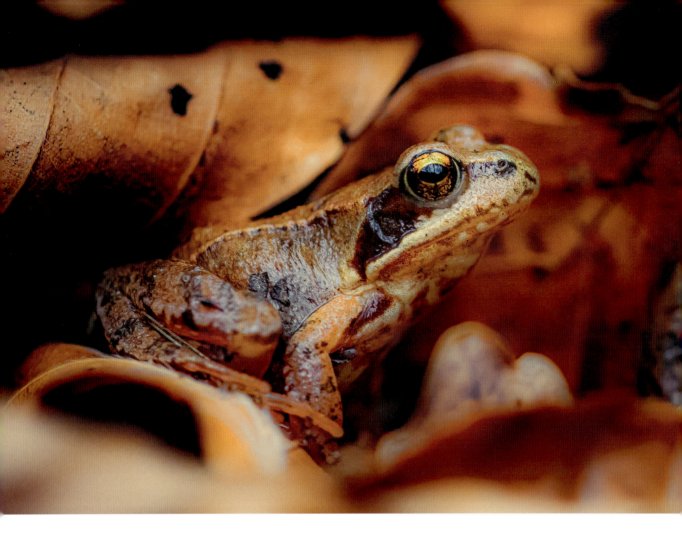

The oldest-known extant clade are archaeobatrachian frogs in the superfamily Leiopelmatoidea, comprising two families—Leiopelmatidae and Ascaphidae—which makes them sister taxa. Intriguingly, leiopelmatid frogs (for example, see Archey's Frog, *Leiopelma archeyi,* page 36) are endemic to New Zealand, while the ascaphid frogs, also known as tailed frogs (see the Rocky Mountain Tailed Frog, *Ascaphus montanus*, page 38, for example), are endemic to western North America. Despite living some 7,456 miles (12,000 km) apart, their morphology and genetics indicate that these two families shared a common ancestor in the Late Triassic (around 237–201 mya). A driver for the divergence of leiopelmatoid frogs is found in the extensive regional volcanic activity that precipitated a mass extinction event, marking the Triassic–Jurassic boundary. The resultant lava flows today cover an area of around 4 million square miles (11 million square kilometers) and would have undoubtedly made this region uninhabitable for frogs for quite some time—certainly long enough for the evolution of now geographically separate independent lineages.

↑ Once metamorphosed, Common Frogs (*Rana temporaria*) spend most of their time on land—in forests and fields, and even in urban parks and gardens.

↗ Arboreality is found across the frog tree of life, and is exemplified in hylid frogs, such as this Italian Treefrog (*Hyla intermedia*). The family Hylidae has a near-global distribution that extends from Europe to the Americas and Australasia.

NEW LIFE TRIGGERED BY GLOBAL EVENTS

For one neobatrachian clade in particular, Hyloidea, another mass extinction event signaled the exploitation of a novel habitat and a new way of life for many frogs: life in the trees. The Chicxulub asteroid impact in the Gulf of Mexico 66 mya wiped out around 75 percent of life on Earth, including the non-avian dinosaurs, and heralded the end of the Cretaceous and the beginning of the Paleogene (the K-Pg boundary).

After the K-Pg event there was a global reduction in forest cover, followed by widescale replacement by ferns and then a subsequent recovery of the forests. Forest recovery seemingly led to a burst of frog diversity and ecological adaptation, as prior to the K-Pg boundary arboreality appears not to have been a life-history mode for anurans. Many hyloid frog families are almost exclusively arboreal (for example, Hylidae and Centrolenidae), and notably an arboreal lifestyle is present in other frog groups (such as the Ranoidea) where it also appeared after the K-Pg boundary. This is an example of convergent evolution—when unrelated organisms evolve similar traits due to similar ecological pressures or opportunities (think of the evolution of flight in birds and bats)—and an illustration of how a novel ecological opportunity was exploited by the diverse and adaptable anurans.

THE MAKING OF MODERN SURVIVORS

CONTINENTAL FRAGMENTS AND ISLAND FROGS

Seychelles sooglossid frogs are the neobatrachian representatives of just two frog families endemic to an archipelago—New Zealand's archaeobatrachian leiopelmatid frogs (for example, Archey's/Hamilton's/Hochstetter's Frog) are the other. Like leiopelmatids, the Sooglossidae (for example, see the Seychelles Frog, *Sooglossus sechellensis*, page 40) have a long evolutionary history, and their present geographic range, although across several islands, is continental in origin.

After Pangaea began to fragment around 200 mya present-day India formed an independent landmass, traveling northward until it collided with Asia and formed the Himalayas about 50 mya. While on this journey, a section broke away and remained isolated in the Indian Ocean, becoming today's Seychelles Inner Islands. Although this paleogeographic history was known, it took intensive molecular studies to confirm where sooglossids sit in the anuran tree of life, and which other frogs are their closest relatives. It is now clear that sooglossids are the sister family to the purple burrowing frogs (Nasikabatrachidae) of India's Western Ghats, providing further evidence that the ancestor of these frogs was present on the larger continental Indian fragment that broke apart around 63 mya.

Seychelles Palm Frog

↑→ Frogs of the island-endemic family Sooglossidae. Seychelles' sooglossid frogs have an ancestry dating back 66 million years. All sooglossids are fully terrestrial, and do not need to enter water at any point in their lives.

250 million years ago

200 million years ago

MODERN FROG DISTRIBUTION

Gardiner's Seychelles Frog

Thomasset's Seychelles Frog

125 million years ago

Global palaeogeography and the appearance of modern frog families

200 mya Pangea begins to fragment—oldest archaeobatrachian frog lineages present; above: 125 mya Laurasia and Gondwana separate—neobatrachian frogs appear and diverge; 66 mya the K-Pg event, India and Seychelles separate—an explosion of neobatrachian frog diversity.

66 million years ago

THE MAKING OF MODERN SURVIVORS

FROGS CROSSING OCEANS

Current distributions of frogs have not all been dictated by large-scale global events. Some pathways are down to chance. Situated between Mozambique and Madagascar lie the Comoros, a group of volcanic islands which, rather than splitting off from a larger landmass, exploded out of the seabed. Every organism living there will therefore have arrived by crossing the ocean, which often happens when animals and plants end up as stowaways in shipped goods or are deliberately transported for aesthetic or economic reasons.

Two species of frogs on the Comoros island of Mayotte were once thought to have been transported there by humans. However, while they are undeniably members of two distinct genera from Madagascar—*Blommersia* and *Boophis*—not only do the Mayotte frogs differ in size and coloration from their Malagasy relatives, they also make different advertisement calls (frog calls are species-specific to prevent misdirected attempts at reproduction with the wrong partner) and are genetically distinct.

It is now thought that *Blommersia transmarina* and *Boophis nauticus* arrived on Mayotte at some point within the last 7 million years, perhaps in two separate events after rafting on vegetation dislodged following heavy rains. Indeed, several other groups of frogs are considered to have historically rafted to new lands by the same method. These include the African reed frogs (family Hyperoliidae)—which have achieved this at least twice, first from Africa to Madagascar with the ancestor of Malagasy reed frogs (genus *Heterixalus*) and then from Madagascar to Seychelles with the ancestor of the Seychelles Treefrog (*Tachycnemis seychellensis*)—and more recently the European Common Toad (*Bufo bufo*, see page 104) which has repeatedly colonized Scottish Islands. Although generally intolerant of salt water, their ability to go for long periods without food and see out adverse conditions (see Chapter 2) means that frogs may be much better colonizers than many mammals.

ARABIAN SEA

The trans-oceanic dispersal of African and Malagasy frogs

Several factors likely contribute to successful transoceanic dispersal in frogs. These include source populations being (i) present at low-elevation coastal areas adjacent to the coast of the new location; (ii) comprised of habitat generalists that are tolerant of sub-optimal conditions, especially while at sea; (iii) able to breed in stagnant water bodies, which can be more common at lower elevations.

SEYCHELLES

COMOROS
MAYOTTE

MADAGASCAR

MOZAMBIQUE CHANNEL

AFRICA

INDIAN OCEAN

MODERN FROG DISTRIBUTION

THE WORLD'S LARGEST AND SMALLEST FROGS

The largest frog in the world is the appropriately named Goliath Frog (*Conraua goliath*) which is found in lowland tropical rivers in western Cameroon and northern Equatorial Guinea (where it was recently rediscovered). Weighing in at around 7 lb 4 oz (3.3 kg) and up to 13 in (34 cm) long, these frogs are the gentle-giant parents of the anuran world. By excavating nests in depressions and gravel banks along their riparian (riverside) habitat, and clearing areas of dead leaves, rocks, and stones, they provide a safe haven for their eggs and tadpoles away from fast flowing water and actively guard their offspring against predators. That Goliath Frogs construct these nests by moving such large objects has been suggested as a selective pressure that favors large body size. Unfortunately, their size makes them a valuable source of protein, and the Goliath Frog is severely threatened by overexploitation from human hunters.

Approximately 40 times smaller than the Goliath Frog, some of the world's smallest frogs are seven species of microhylid in the genus *Paedophryne* from Papua New Guinea. These diminutive frogs live in the damp leaf litter of upland forests, the smallest of which, the fly-sized *P. amauensis*, averages just 0.31 in (7.7 mm) long, which until recently made it not only the world's smallest frog, but also the world's smallest vertebrate. However, while writing this book, an even smaller frog was confirmed. *Brachycephalus pulex*, known only from a single site in Brazil's Atlantic Rainforest, averages just 0.25 in (7 mm).

The greater a frog's body size, the smaller the surface area of skin there is relative to body volume. This means that smaller frogs have a higher surface area over which to lose water, and so are more prone to desiccation. They therefore tend to be found in tropical forests at higher elevations, where their microhabitat remains stable. Although the reproductive behavior of *Paedophryne* frogs is unknown, many microhylids are direct developers, laying eggs on land that hatch out as froglets. In doing so, they avoid exposing their tadpoles to the abundant invertebrate predators present in water bodies. However, the evolution of this life-history trait, which has appeared independently across several anuran families, may also be an adaptation to elevated (mountainous) terrain and the general absence of water bodies in such habitats.

↑ Only males of the frog *Paedophryne amauensis* (top) have ever been found. Despite being the world's largest frog (bottom), the Goliath Frog has a relatively small range in Central Africa spanning southwestern Cameroon and northwestern Equatorial Guinea.

Paedophryne amauensis
0.31 in (7.7 mm)

American Toad
3.5 in (9 cm)

European Common Toad
6 in (15 cm)

Goliath Frog
13.5 in (34 cm)

To scale

From the largest to the smallest
Scaled representations of the Goliath Frog, European Common Frog, American Toad, *Paedophryne amauensis*.

THE MAKING OF MODERN SURVIVORS

RANA TEMPORARIA

European Common Frog

One of the first amphibians described by Carl Linnaeus

SCIENTIFIC NAME:	Rana temporaria
FAMILY:	Ranidae
LENGTH:	2½–3½ in (65–90 mm)
LIFE HISTORY:	Aquatic eggs (up to 4,500 in clumps) and tadpoles; terrestrial adults
NOTABLE FEATURE:	To reject unwanted males, female frogs may rotate their body, imitate male calls, and even feign death
IUCN RED LIST:	Least Concern

A relatively large "true" frog, *Rana temporaria* has the typical appearance of a frog, and the "standard" natural history of temperate anurans: springtime emergence from hibernation for a period of explosive breeding in a water body, resulting in gelatinous, floating egg masses and tadpoles undergoing aquatic development before they leave the water several weeks later. European Common Frogs will hibernate on land or in water, and many hibernate in flowing water, which may reduce the risk of freezing.

Found from Eastern Europe to the British Isles in lowland grassland, farmland, parks, gardens, and highland forests, the European Common Frog even ranges north of the Arctic Circle into tundra habitat. The digestive tract of common frogs often partially degenerates during hibernation, and its regeneration seemingly corresponds with the lack of feeding (at least in males) during the reproductive period, with the frogs relying on stored reserves while regeneration takes place. When frogs do begin to feed, energy is used for growth and tissue repair, but toward autumn and into winter their metabolism progressively directs energy to fat stores to see them through hibernation.

Perfectly camouflaged for terrestrial life with its variable brown to olive, reddish, and yellow coloration, the European Common Frog bears a strong resemblance to the North American Wood Frog (*Lithobates sylvaticus*, see page 72), including in ecology and behavior, with frogs in warmer parts of their range breeding earlier in the year and those in cooler regions breeding later. This pattern is reflected in a positive relationship between elevation and breeding period. However, although the European Common Frog can withstand freezing temperatures, it can only do so for short periods.

→ After metamorphosing from a tadpole, European Common Frogs spend the next few years under cover in scrublands, parks, gardens, fields, and woodland, feeding and growing until they reach adult size and are ready to breed.

THE MAKING OF MODERN SURVIVORS

ANAXYRUS AMERICANUS

American Toad

One of the most widely distributed North American anurans

SCIENTIFIC NAME:	*Anaxyrus americanus*
FAMILY:	Bufonidae
LENGTH:	2¼–4½ in (55–110 mm)
LIFE HISTORY:	Aquatic eggs (up to 8,000 in strings) and tadpoles; terrestrial adults
NOTABLE FEATURE:	Female toads avoid mating with close relations, possibly via odor and call recognition
IUCN RED LIST:	Least Concern

Ranging from Texas to as far north as Newfoundland and Labrador, the American Toad (a "true" toad) is one of the most widely distributed North American anurans, and the most widely distributed bufonid in North America. Looking like a "typical" toad, with a robust body, warty-looking, granular skin, short legs, and walking gait, American Toads can have a yellowish, olive, or brown general coloration, often dotted with dark to light brown spots and with a light-colored line stretching dorsally from the head. As in other toads, American Toads possess large parotoid glands behind the head which contain toxic secretions (bufotoxin) as a defence against predators.

Anaxyrus americanus is sometimes referred to as a species group, comprising three subspecies—the Eastern American Toad (*A. a. americanus*), the Dwarf American Toad (*A. a. charlesmithi*), and the Hudson Bay Toad (*A. a. copei*)— although there is regular hybridization between certain populations. The American Toad appears tolerant of disturbance and is found in a variety of urban and human-managed environments, including scrub, woodland, forests, grassland, wetlands, farmland, parks, and gardens. Perhaps due to their experience when initially developing as larvae (eggs of the same clutch being in close proximity), the tadpoles of the American Toad are able to recognize their near relatives and often form large aggregations, likely as a defence against predation. Similarly, newly metamorphosed toadlets often congregate along the margins of their natal ponds prior to dispersal en masse to dense vegetation.

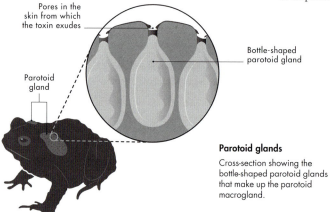

Parotoid glands
Cross-section showing the bottle-shaped parotoid glands that make up the parotoid macrogland.

→ With a relatively short but stout body, warty granular skin, and large parotoid glands, the American Toad is very much a "typical" toad from the family Bufonidae.

THE MAKING OF MODERN SURVIVORS

LEIOPELMA ARCHEYI

Archey's Frog

At more than 35 years, this is one of the longest-lived frogs in the world

SCIENTIFIC NAME:	*Leiopelma archeyi*
FAMILY:	Leiopelmatidae
LENGTH:	Up to 1½ in (40 mm)
LIFE HISTORY:	Terrestrial eggs (4–19), tadpoles (carried by the male to complete development) and adults
NOTABLE FEATURE:	With no mating calls, these frogs are instead thought to communicate using chemical signals
IUCN RED LIST:	Critically Endangered

Archey's Frog is the smallest of three species of archaeobatrachian frogs within the genus *Leiopelma* (the other species are Hochstetter's Frog, *L. hochstetteri*, and Hamilton's Frog, *L. hamiltoni*). Leiopelmatids are the only New Zealand endemic frogs, and one of only two anuran families endemic to an archipelago (the other being Seychelles' Sooglossidae). They have a greenish-brown background coloration, but their scattered dark lines and blotches are unique to each frog and can be used to identify individuals.

Several characteristics make Archey's Frog (and the other leiopelmatids) highly distinct compared to most other frogs, and illustrate their long evolutionary history. These features include the presence of nine vertebrae, ventral inscriptional (cartilage, not bone) ribs, and vestigial tail-waggling muscles that are retained after the tadpoles metamorphose. With metamorphosis taking three to four years to complete, the remnant tail musculature provides extra surface area through which the young frogs can breathe until their lungs are fully formed. Archey's Frog is entirely terrestrial and spends its life in the native forests on North Island, secreted under stones, logs, and vegetation on the forest floor, emerging to feed at night when rainfall and humidity present ideal conditions for foraging and reproduction. Along with the other leiopelmatid frogs, Archey's Frog is of conservation concern due to habitat loss, introduced predators, and disease, and is subject to ongoing initiatives to protect and increase the remaining populations.

→ Archey's Frog is the smallest of the four species of leiopelmatid frog, and the world's most Evolutionarily Distinct and Globally Endangered (EDGE) amphibian.

THE MAKING OF MODERN SURVIVORS

ASCAPHUS MONTANUS

Rocky Mountain Tailed Frog

A North American endemic family

SCIENTIFIC NAME:	*Ascaphus montanus*
FAMILY:	Ascaphidae
LENGTH	1–2 in (30–50 mm)
LIFE HISTORY	Aquatic eggs (40–75 in strings) and tadpoles; aquatic/semi-terrestrial adults
NOTABLE FEATURE	Only frog with a copulatory organ used for internal fertilization
IUCN RED LIST	Least Concern

Alongside their sister taxon—the leiopelmatid frogs of New Zealand from which they are estimated to have diverged around 200 mya—ascaphid frogs are among the oldest known extant anuran lineages, being remarkably similar to some of the earliest frogs in elements of their skeletal morphology (for example, nine pre-sacral vertebrae, as in *Callobatrachus*, whereas all other extant frog taxa generally possess eight).

While several frog species have internal fertilization, the Rocky Mountain Tailed Frog and its sister species, the Coastal Tailed Frog (*Ascaphus truei*), are the only anurans with a copulatory organ—the "tail" observed in the males. The "tail" of both species is an extension of the cloaca (the combined excretory and genital opening). This is thought to be an adaptation to living (and breeding) in fast-flowing streams, where these frogs can be found in high-gradient, mid-elevation montane areas, usually in mature forests. Mating in *A. montanus* may last up to 90 hours, and the females are able to store sperm, meaning that mating and egg deposition can occur at different times of year, perhaps nine months apart.

Tadpoles of the Rocky Mountain Tailed Frog have suctorial mouthparts allowing them to cling to rocks in moving water. The mouthparts may even allow them to emerge from the water onto rocks at night, possibly to graze on algae. Tadpole development can take up to three years, and following metamorphosis, a further seven to eight years until frogs are able to breed, with females thought to reproduce every other year.

As fast-flowing water is highly oxygenated, the Rocky Mountain Tailed Frog breathes mainly through the skin rather than the lungs. Frogs are generally brown in color, with darker patches and mottling on their granular skin.

→ As well as continent-scale geographic events in the Late Triassic that likely drove lineage divergence in the ancestor of *Ascaphus* and *Leiopelma*, a more recent and localized event is thought to have isolated ancestral *Ascaphus* frogs, leading to the two species we see today, the Rocky Mountain Tailed Frog and the Coastal Tailed Frog (*Ascaphus truei*).

THE MAKING OF MODERN SURVIVORS

SOOGLOSSUS SECHELLENSIS

Seychelles Frog

An ancestral lineage that dates back 66 million years

SCIENTIFIC NAME:	*Sooglossus sechellensis*
FAMILY:	Sooglossidae
LENGTH:	½–¾ in (15–21 mm)
LIFE HISTORY:	Terrestrial eggs (5–15 in clumps), tadpoles, and adults; parental care
NOTABLE FEATURE:	Female cares for the young tadpoles and froglets by carrying them on her back (they only feed after they have left her)
IUCN RED LIST:	Endangered

The Sooglossidae are one of only two anuran families endemic to an archipelago (the other being New Zealand's Leiopelmatidae). They are among the oldest known neobatrachian lineages, having diverged from their closest living relative, India's purple burrowing frogs (Nasikabatrachidae) 66 mya—meaning these frogs survived the asteroid impact that wiped out the dinosaurs. The Seychelles Frog is found on three of the granitic inner islands—comprised of several mountain peaks sitting atop a microcontinental fragment of Gondwana (most of which is submerged around 180 ft/55 m below sea level)—and like other sooglossids has an entirely terrestrial life history.

Despite having no external or middle ear, the Seychelles Frog makes complex vocalizations with at least two different notes—the majority of frogs repeat just a single note. The sound is possibly received through bone conduction in the head and/or forearms and shoulder, vibrations through the body wall, and via the lungs toward the inner ear, or via the opercularis system (unique to amphibians and considered an adaptation to life on land). To attract females, male frogs vocalize from cool, damp patches beneath leaf litter or from cracks and crevices in tree roots or rocks. Eggs are deposited, fertilized, and then guarded by one or possibly both parents. Larvae develop into tadpoles within the egg, and on hatching climb on to the back of the female, where they remain until they metamorphose into froglets just a few millimeters long. Throughout this time, the young do not feed, subsisting solely on the yolk.

Being colored a mixture of browns, and sometimes even pale pink, with dark patches and bands, dotted with orange-tipped tubercles, and often with small patches and dots of blue, the Seychelles Frog perfectly blends into its forest-floor habitat, where it is found in mid- to high-elevation mist, palm, and mixed forest, often adjacent to streams. There is evidence to suggest that these frogs have adapted to historic climate warming, as those from Praslin (1,204 ft/367 m elevation) are found in much lower and drier habitats than frogs from either Silhouette (2,248 ft/740 m) or Mahé (2,969 ft/905 m) in the Seychelles.

→ The Seychelles Frog has a distinct genetic identity that is specific to each of the three islands—Mahé, Silhouette, and Praslin—where it is found. Frogs from Praslin may be the oldest of these three lineages, at an estimated 5.7 million years.

LIFE ON LAND
& IN WATER

LIFE ON LAND AND IN WATER

Living in plants

In areas where ponds and other water bodies are few and far between, but vegetation is abundant, many frogs make use of the smaller collections of water often found in plants. They use these microhabitats both as a refuge and a place to breed and care for their young.

Bromeliads are flowering plants found mainly in the Neotropics. Many are rootless and grow on trees, using leaf axils to collect rainwater and nutrients. A range of frogs from different families use bromeliads as temporary refuges, and around 100 species complete their entire life cycle within these plants. This arrangement may benefit both frogs and the bromeliad: the eggs hatch into tadpoles which can grow in a relatively benign environment free of predators, while the bromeliad gains nutrients from tadpole feces.

← Many frogs, like this slender-legged treefrog (genus *Osteocephalus*) deposit their eggs in water captured within the leaves of tropical plants—phytotelma—which act as nurseries for their developing young.

→ Pitcher plants hold rainwater, and many such reservoirs are used by frogs to provide a stable and secure habitat for their developing tadpoles, like those of the Matang Narrow-mouthed Frog (*Microhyla nepenthicola*).

Many such frogs are tied to specific bromeliad species and have highly restricted distributions. For example, the Itambé Bromeliad Frog (*Crossodactylodes itambe*) is found in only one species of bromeliad that occurs on a single mountain summit in Brazil, and the entire global range of the frog is less than a quarter of a square mile. Dispersal and colonization of new bromeliads is thought to occur after heavy rain, although neither this behavior, nor any life history stages of the Itambé Bromeliad Frog, have ever been observed outside its host plant.

Frogs also breed in other water-holding plants like the pitcher plants. Characteristic of poor soils, these plants increase their nutrient intake by capturing and digesting insects within a leafy, water-holding vessel, or "pitcher." Visual cues, scent, or nectar released by the pitchers attract insects, and a slippery pitcher rim, waxy inner walls, and inward-facing, bar- or hair-like structures prevent them escaping, after which the insects drown and are digested in the fluid-filled pitcher. In North America, Spring Peeper Frogs (*Pseudacris crucifer*) sometimes use pitcher plants as refuges, and benefit by intercepting and consuming insects attracted to the plants.

Being mainly carnivorous, pitcher plants release specific organic chemicals into the pitcher that digest animal matter, making their use as tadpole nurseries hazardous. However, the Borneo Narrow-mouthed Frog (*Microhyla borneensis*)—among the world's smallest at less than ½ in (13 mm)—does use a pitcher plant (*Nepenthes ampullaria*) in which to lay eggs and raise its tadpoles. This relationship is noteworthy, as the frog seemingly exploits the fact that *N. ampullaria* is primarily a detritivore, relying on dead leaves for its nutrients, and therefore provides a relatively safe environment for the frog's tadpoles to grow and develop.

LIFE ON LAND AND IN WATER

Up in the trees

Living above the ground in trees and bushes (arboreality) has evolved in at least 17 frog families, with three (Centrolenidae, Hylidae, Rachophoridae) comprising almost exclusively arboreal species. This is a good example of convergent evolution, where the same trait evolves separately in different groups, and demonstrates how trees provide a range of useful resources that frogs can exploit.

Among treefrogs, there are a variety of specialisms, with some using vegetation several feet above ground or exploiting tree holes for refuges and breeding, while others have mastered the high canopy but return to lower vegetation to lay eggs in cavities, ground-level pools, foam nests, or on leaves that then drip developing tadpoles into a water body. For example, among the three species of Central American hylid treefrogs within the genus *Cruziohyla*, all deposit eggs overhanging pools of water, with the tadpoles dropping into the water below upon hatching. However, while both the Fringed Leaf Frog (*C. craspedopus*) and Sylvia's Treefrog (*C. sylviae*) lay their eggs immediately above flooded cavities or water-filled hollows in fallen trees, the Splendid Treefrog (*C. calcarifer*) lays its eggs on leaves overhanging larger terrestrial pools and open water bodies.

Living in trees and bushes and moving through swaying branches and foliage, presents challenges for frog locomotion, especially in windy conditions. Consequently, sticky toe pads that allow the frogs to adhere to the substrate and climb smooth surfaces have evolved independently in several anuran families.

↗ The Waxy Monkey Frog (*Phyllomedusa sauvagii*) from South America lays its eggs in a nest constructed from leaves.

← The Fringed Treefrog (*Cruziohyla craspedopus*) from South America lives high in the forest canopy, but descends to the forest floor to breed in small pools.

UP IN THE TREES

Foot adaptation
The feet of frogs are highly adapted to their ecology and behavior, from webbed aquatic and semiaquatic, to more robust terrestrial walking and burrowing, arboreal walking, and even gliding.

Mwanza Frog
(*Xenopus victorianus*)
Aquatic

Telmatobius chusmisensis
Aquatic

Moyer's River Frog
(*Amietia moyerorum*)
Aquatic

Red-bellied Toad
(*Melanophryniscus dorsalis*)
Terrestrial walker

Eastern Spadefoot
(*Scaphiopus holbrookii*)
Terrestrial burrower

Pseudopaludicola pocoto
Terrestrial/semiaquatic

Lemur Leaf Frog
(*Agalychnis lemur*)
Arboreal walker

Fringe-limbed treefrog
(*Ecnomiohyla miliaria*)
Arboreal glider

Treefrogs like the American Green Treefrog (*Dryophytes cinereus*, see page 64) can comfortably attach to a vertical glass surface without sliding off, even when this is tilted at an acute angle. Microscopic examination of the toe pads has revealed a complex structure that enables treefrogs to maintain their grip. The pads contain hexagonal-shaped cells separated by spaces filled with mucus, which provides a degree of "wet adhesion" and can increase the friction between the toe pad and the substrate. Other methods of hanging on include opposable digits that can grip a branch or leaf, rather like the hand of a primate, as is the case with frogs in the genus *Phyllomedusa*, also known as monkey treefrogs.

Along with gripping onto vertical surfaces, leaves, and branches, arboreal frogs must still traverse their environment to feed and find a mate. Many do this by conventional jumping, walking, or climbing. The Asian treefrogs (or gliding frogs) comprise over 400 species in 21 genera. They are characterized by large webbing on the feet and skin flaps that enable them to glide down from an arboreal vantage point (see Chapter 5).

LIFE ON LAND AND IN WATER

Living in water

Frogs have successfully exploited a wide range of water bodies, for breeding in particular. These include ephemeral ponds, which cyclically (often seasonally) fill and then dry out, and more expansive habitats such as lakes. Many anurans also live and breed in flowing water like rivers and streams and have evolved physical and behavioral adaptations to thrive in these environments.

TEMPORARY PONDS

The habitats that frogs have been most successful in exploiting, especially for breeding, are temporary ponds, which fill during the rainy season or after snowmelt, and then slowly desiccate when dry conditions prevail (these are sometimes also called "vernal pools"). These ephemeral water bodies can range in size from small tractor ruts on hillside tracks to large, shallow ponds that form in the depressions between sand dunes. As they are used primarily for breeding and only for a relatively short period, many frogs do not feed while in the water. For example, the "true" frogs and toads in the families Ranidae and Bufonidae mostly prey on terrestrially captured invertebrates. Consequently, they may not forage at all during the breeding period, although this may be over in just a few days.

→ The European green frogs (genus *Pelophylax*) are a group of highly aquatic species that often sit immobile in the water with just their head visible, waiting to ambush insects that may alight on the surface, or to dive down into the water to escape from predators.

LIFE ON LAND AND IN WATER

Temporary ponds are usually unsuitable for fish, which would be left high and dry when the pond dries out. Although not widely used by adult frogs as places to feed, rain-fed ponds are rapidly colonized by microorganisms, invertebrates, algae, and plants, providing a rich food supply for developing tadpoles. The challenge then for the tadpoles is to complete their development and transform into a froglet before the pool dries out. Frogs deal with this challenge in several ways. One approach is to deposit eggs across different pools. Although some pools—perhaps the majority—may dry out, with the resulting loss of all the young, others might retain water long enough for tadpoles to complete their metamorphosis.

Another strategy (often combined with depositing eggs across different pools) is to breed repeatedly over several years. This requires adult frogs to live long enough to have multiple breeding opportunities throughout their lives. If females lay eggs every year for five years, all the young could be lost due to pond

LIVING IN WATER

↖ The Moor Frog (*Rana arvalis*) breeds in boggy pools in northern Europe, and the males can turn bright blue during the mating season.

↗ When frogs like this European Common Frog (*Rana temporaria*) metamorphose, aside from a remnant tail stub and a disproportionately large head, they emerge from the water as a miniature version of the adult.

desiccation in four of those years, but just one successful year may be sufficient to produce enough froglets that survive to adulthood, and thereby sustain the population.

Some tadpoles can even accelerate their development in response to pond desiccation as well as other threats (see Chapter 4). An alternative to laying multiple clutches in different water bodies, and/or over several years, is for the adults to produce fewer offspring but invest more time and effort in caring for them (see also Chapter 4).

PERMANENT WATER BODIES

Permanent and semi-permanent water bodies, such as lakes, reservoirs, spring-fed tarns, or those that form in disused quarries, present different challenges for frogs. Although desiccation risk is reduced, these habitats are much more likely to contain predators. Some frog species may therefore confine their activities to shallower or well-vegetated areas. Alternatively, the tadpoles of those frogs that breed in deeper water often have skin toxins, which make them unpalatable to fish and other potential predators (see Chapter 7).

Frogs of the family Pipidae—which comprises African Clawed Frogs and the Surinam Toads of South America—spend almost their entire lives in more permanent ponds and slow-moving streams, so they do feed in the water. The African Clawed Frogs, for example, have smooth skins that are equipped with numerous organs that superficially look like stitches in the skin. These are lateral line organs—similar to those possessed by fish—which can detect the movements of both predator and prey in the water.

↑ Permanent water bodies often also host fish and waterfowl that may make a meal of frogs, tadpoles, and eggs.

Although all aquatic/semi-aquatic frogs have webbed hind feet, these are particularly well-developed in the pipid frogs and provide extra thrust when swimming. These are the only frogs to have claws (or, strictly speaking, claw-like structures) on their hind feet. Most anurans only take live, moving prey, but clawed frogs also take carrion, using their claws to tear apart a larger item of food by holding it in their mouths and bringing their hind legs forward in a kicking motion. The front legs are often used for feeding, pushing food items into the mouth.

As they spend much of the time either buried in mud or suspended at the surface of the water, the eyes of clawed frogs are positioned on the top of the head, enabling them to look out for potential predators (herons and egrets, for example) that may be stalking the pond or river. Although very much adapted to a life in water, even clawed frogs will move short distances across land if they are disturbed or should their pond begin to dry out.

The family Telmatobiidae, endemic to the South American Andes, contains over 60 species of highly aquatic frogs. All known taxa belong to the genus *Telmatobius*, which has two of the largest totally aquatic frogs in the world, the Andes Smooth Frog or Lake Junin Frog (*Telmatobius macrostomus*) and the Titicaca Water Frog (*T. culeus*, see page 66), both of which are found in high-elevation lakes.

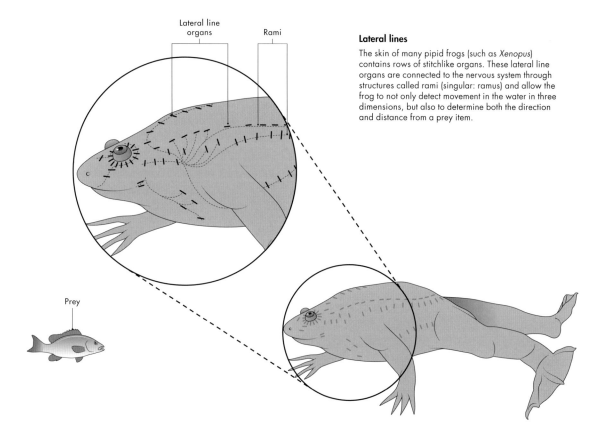

Lateral lines

The skin of many pipid frogs (such as *Xenopus*) contains rows of stitchlike organs. These lateral line organs are connected to the nervous system through structures called rami (singular: ramus) and allow the frog to not only detect movement in the water in three dimensions, but also to determine both the direction and distance from a prey item.

FLOWING WATER

Depending on the volume and seasonality of rainfall, streams and rivers can be permanent, or flow for just a short time in a given year. Breeding in these environments presents another challenge to frogs, but several species have adapted their behavior and morphology by anchoring eggs under rocks or within vegetation and evolving tadpoles with suctorial mouthparts that enable them to cling to rocks (see Chapter 4). Additionally, tadpoles adapted to flowing water usually have more streamlined bodies and stronger tail muscles in order to swim within a current.

↑ Frogs that breed in flowing water may have streamlined tadpoles with strong tail muscles for swimming, as well as suckers that allow them to cling to rocks.

→ The Sabah Huia Frog (*Huia cavitympanum*) is also known as the "hole-in-the-head" frog on account of the tympanic membrane which, unlike other frogs, sits in a recess behind the eye. This adaptation enables it to better detect the ultrasonic calls made by this species.

Alternatively, some species will avoid using the main torrent or river and instead utilize backwaters, or temporary pools that form along the river corridor in times of heavy rain or flood.

Although fast-flowing water presents some challenges for frogs, the constant churning and aeration of the water means that such torrents are usually well-oxygenated. Until 2007, only two specimens of the Bornean Flat-headed Frog (*Barbourula kalimantanensis*) were known to science, and these had been collected from fast-flowing rivers in the Indonesian state of Kalimantan. Subsequent surveys revealed the existence of two further populations upstream from the original localities, and a recent analysis of the specimens revealed that this species has only tiny remnant lungs. Its flattened body not only helps this aquatic frog to hide in tight spaces between and under rocks, but also provides sufficient surface area over which to absorb oxygen from the turbulent waters where it lives.

HEARING FROG CALLS

Frogs tend to be vocal amphibians and display a complex range of audible calls to defend their territories and to attract mates (see Chapter 3). The challenge with breeding in the vicinity of turbulent streams or waterfalls is that the cascading water creates a complex bioacoustic environment which may mask frog calls. This is particularly the case when males form a breeding chorus against a loud backdrop of running water. Several species that breed in such habitats use visual signals to complement their calls (see Chapter 3). Two stream-dwelling species—the Concave-eared Torrent Frog (*Odorrana tormota*) from China and the Sabah Huia Frog (*Huia cavitympanum*) from Borneo—emit ultrasound calls (beyond the range of human hearing and much like many bats) that are detectable by other frogs above the lower frequency sound of rushing water.

Many parts of the Amazonian rainforest are subject to seasonal flooding which can cause dramatic changes in river levels. Mats of floating vegetation often break off and create what are known as "floating meadows" within such flooded systems. Hylid treefrogs are abundant in these floating meadows and may breed within them. Frogs from other families (for example, Bufonidae, Leptodactylidae) are also often present, but as such moving habitats can travel long distances downstream, they are probably also useful for dispersal, and may explain the distributions of some of the species that use them.

1 Surinam Toad 6.7 in (17 cm)
2 Slender-fingered Toadlet 2 in (5 cm)
3 Henle's Snouted Treefrog 1.2 in (3 cm)
4 Tarauaca Snouted Treefrog 1.2 in (3 cm)
5 Orinoco Lime Treefrog 1.6 in (4 cm)
6 Cane Toad 6 in (15 cm)
7 Basin Treefrog 2.4 in (6 cm)
8 Harald's Treefrog 0.8 in (2 cm)
9 Manaus Slender-legged Treefrog 3.2 in (8 cm)

Floating meadow

Floating meadows are beds of vegetation that occur on rivers in Amazonia. They contain distinct assemblages of frogs that use different parts of the vegetation. The measurements alongside the species listed to the left indicate body lengths.

LIFE ON LAND AND IN WATER

Challenges of life on land

To avoid the hottest, driest, and/or coldest periods, many anurans spend hours, days or months secreted in cracks and crevices, hidden on the underside of leaves, or buried in leaf litter, soil, or at the bottom of a pond. They emerge only when conditions are right for them to be active, feed, and reproduce. These behaviors and associated adaptations have allowed frogs to exploit some of the most extreme environments on Earth.

Although all frogs have an association with water and can be found in habitats ranging from tiny puddles to large lakes, the majority spend most of their lives on land. In fact, many frogs do not need freestanding water in which to breed at all. The Coquí Frog (*Eleutherodactylus coqui*) from Puerto Rica, for example, deposits its eggs in damp places. The tadpoles then complete their development within the egg capsule and hatch as fully metamorphosed froglets.

The terrestrial activities of many frogs means that high temperatures present a very real risk of desiccation. In response, frogs will use climate-stable refugia such as tight spaces under rocks and logs or burrows in which to retreat,

← Most frogs in the genus *Eleutherodactylus* lay eggs on land which hatch as fully developed froglets.

↗ In the northern part of their North American range, Eastern Spadefoot Toads (*Scaphiopus holbrookii*) hibernate during winter. In the milder south, they stay active year-round.

emerging only on cool, moist nights to forage. The evolution of such behaviors has allowed frogs to colonize and adapt to seasonal extremes of hot and cold conditions. Couch's Spadefoot Toad (*Scaphiopus couchii*, see page 68) is a well-studied North American anuran with an intriguing suite of adaptations to dry conditions, many of which are convergent with desert-adapted frogs of Africa and Australia, such as the Northern Burrowing Frog (*Neobatrachus aquilonius*, see page 70).

Maintaining hydration when on land is crucial for frogs. Water can be absorbed through the skin, and particularly through the underside when in direct contact with a damp substrate. A patch of skin on the belly in front of the hind legs, sometimes known as the "drink patch" or "pelvic patch," has a rich supply of blood vessels and can absorb water very effectively.

LIFE ON LAND AND IN WATER

↑ The Australian Water-holding Frog (*Cyclorana platycephala*) is well-adapted to survive dry conditions, and stimulated by rainfall to emerge from their protective burrows to breed.

ESTIVATION AND HIBERNATION

To survive dry desert conditions but remain hydrated, many anurans—for example, Australia's Water-holding Frog (*Cyclorana platycephala*)—store water in their bladder and form waterproof cocoons from layers of shed skin. By estivating in this way, Water-holding Frogs can remain buried for up to three years at a time, and as a result have provided a useful source of water for indigenous peoples in times of drought.

Frogs in more temperate climates avoid colder periods by hibernating, but some go one step further by physiologically adapting to extreme cold, while several North American species can even tolerate freezing (for example, some species of North American chorus frogs, *Pseudacris crucifer*, *P. triseriata*, and *P. regilla*, and the hylid treefrogs, *Dryophytes versicolor* and *D. chrysoscelis*). However, one frog truly is a cold-habitat survival specialist, and ranges farther north into the Arctic Circle than any other anuran. The Wood Frog (*Lithobates sylvatica*, see page 72), which is found across North America including Alaska and Canada, uses a form of natural antifreeze to survive subzero temperatures during the winter months.

Gray Treefrog (Elevation 67 ft/20.5 m)

Asian Black-webbed Treefrog (Elevation 190 ft/57 m)

Titicaca Water Frog (Depth 400 ft/120 m)

High Himalayan Frog (Elevation 15,000 ft/4,500 m)

Frogs high and deep
The High Himalaya Frog (*Nanorana parkeri*) is the only known amphibian to exist at such high elevations— around 15,000 ft (4,500 m) on the Tibetan Plateau. The Titicaca Water Frog (*Telmatobius culeus*) is the deepest known diver, in Lake Titicaca, South America. The Asian Black-webbed Treefrog (*Rhacophorus kio*) is currently the highest known climber.

CHALLENGES OF LIFE ON LAND

↑ Wood Frogs (*Lithobates sylvaticus*) can survive the extremely cold North American winters by using a natural antifreeze within their tissues.

→ The breeding period of the Pacific Treefrog (*Pseudacris regilla*) not only varies considerably in response to prevailing climatic conditions across its western North American range, but also according to elevation, being found up to 10,000 ft (3,048 m) above sea-level.

LIFE ON LAND AND IN WATER

MANAGING EXTREME CONDITIONS

While frogs have a body temperature that varies according to their immediate thermal environment, which they can manage through behavioral thermoregulation (in other words, they are ectothermic), mammals and birds can control their body temperature internally through their metabolism (they are endothermic). Mammals and birds can therefore remain active in a variety of thermal environments, but anuran activity and metabolism slow under colder conditions, and frogs usually need to avoid extremes of temperature. They do this by seeking a refuge—deep leaf litter, a burrow, the bottom of a deep pond or lake—to wait out the cold and/or hot extremes.

Hibernation infers a period of dormancy in seasonally cold conditions, but as the physiological responses to cold are different between endotherms and ectotherms the term "brumation" is often applied to frogs and other ectotherms. Conversely, the avoidance of seasonally hot, dry conditions is referred to as estivation, and applied more generally across the animal kingdom.

Survival Strategies: Effects and Benefits

As anuran growth slows during hibernation and/or estivation, distinct "lines of arrested growth" are laid down in the bones rather like the growth rings of trees. By counting these growth rings, it is possible to calculate the number of winters (or hot summers) that a frog has survived, and therefore estimate the animal's age.

The endothermic state in mammals and birds means they can remain active for longer during variable thermal conditions but require regular food to maintain such activity. By synchronizing periods of reduced activity with extremes of thermal environment, frogs can divert the energy otherwise needed to maintain a constant body temperature to reproduction. Along with other ectotherms (like other amphibians, reptiles, fishes, and invertebrates), frogs therefore need less food than mammals and birds, and can often reproduce faster and generate greater numbers of offspring during their periods of peak activity.

→ The Marbled Snout-burrower (*Hemisus marmoratus*) is a subterranean burrowing frog found across sub-Saharan Africa.

↗↗ The Water-holding Frog (*Cyclorana platycaphela*) of Australia survives dry desert conditions by burrowing into the soil and forming a cocoon from shed skin.

→→ Found in France, Spain, and Portugal, in the northern part of its range, the Western Spadefoot Toad (*Pelobates cultripes*) will hibernate in a self-dug burrow to escape colder periods. In hotter, southern regions it burrows in order to estivate.

CHALLENGES OF LIFE ON LAND

LIFE ON LAND AND IN WATER

DRYOPHYTES CINEREUS

American Green Treefrog

Life in the trees

SCIENTIFIC NAME:	*Dryophytes cinereus*
FAMILY:	Hylidae
LENGTH:	1½–2½ in (30–60 mm)
LIFE HISTORY:	Aquatic eggs (400–2,500 in clumps) and tadpoles; terrestrial arboreal adults
NOTABLE FEATURE:	The adhesion of this treefrog's feet can withstand a force of more than 0.6 kg
IUCN RED LIST:	Least Concern

As its common name suggests, the American Green Treefrog is a highly arboreal species native to the United States. It inhabits wooded open canopy areas in the vicinity of lakes, ponds, marshland, and backwaters. Coloration varies from bright through to dark green, yellow, brown, and sometimes slate-gray. Most individuals have a yellow, cream, or light-colored lateral stripe, and like many anurans, they can both darken and lighten their skin to match the background habitat.

The thermal ecology of the American Green Treefrog means that it can maintain a higher body temperature than non-arboreal frogs: as much as 97°F (36°C) when in direct sunlight. This is likely due to the production of a mucus barrier, which helps the frog to limit evaporative water loss. Most amphibians have a low tolerance of saline conditions, but intriguingly, coastal populations of American Green Treefrogs have been found breeding in brackish water. The mucus barrier may contribute to this salt tolerance, but the coastal populations also express genes for salt tolerance that are missing in the inland freshwater populations. A further adaptation to such conditions is a shortened larval period, which reduces the time that tadpoles are exposed to high salt levels during their development.

This species is currently expanding its range in the United States, driven primarily by increased flooding due to climate change. Frogs in the expanded range have longer hind legs—indicating greater dispersal ability—than frogs in the historic range.

→ The American Green Treefrog is a chorus breeding frog, meaning that many individuals, male frogs in particular, will congregate at a breeding site to attract a mate. Male frogs will also physically compete with each other by butting or wrestling.

LIFE ON LAND AND IN WATER

TELMATOBIUS CULEUS

Titicaca Water Frog

Frog with wrinkles

SCIENTIFIC NAME:	*Telmatobius culeus*
FAMILY:	Telmatobatidae
LENGTH	3–5½ in (75–140 mm)
LIFE HISTORY	Aquatic eggs (500 in clumps), tadpoles, and adults
NOTABLE FEATURE	The skin is thrown into a series of fleshy folds resulting in its alternative name of "scrotum frog"
IUCN RED LIST	Endangered

The Titicaca Water Frog is a large aquatic frog found in Lake Titicaca and adjacent water bodies on the border between Bolivia and Peru. It has a background coloration ranging from gray to brown and olive-green, with darker mottling or marbled patterning. When stressed, this frog exudes a sticky, milk-like secretion that fills its skin folds.

The fully aquatic nature of the Titicaca Water Frog is likely due to the extreme environmental conditions surrounding the lake. Situated 12,500 ft (3,800 m) above sea level, Lake Titicaca is subject to strong winds, high UV exposure, and temperatures of 19–64°F (-7–18°C). Temperature fluctuations are much reduced within the lake, and the frog is generally observed within 10 ft (3 m) of the surface where conditions are a more stable 52–57°F (11–14°C), although it has been recorded at depths of 394 ft (120 m) and may use the entire lake water column of 955 ft (291 m).

The Titicaca Water Frog has much-reduced lungs, instead absorbing most of its oxygen across the fleshy folds along its body and legs. These skin folds are richly innervated with blood vessels and maximize the skin surface area for oxygen absorption. In addition, this frog has a much higher count of oxygen-carrying red blood cells than other amphibians and also one of the lowest metabolic rates of any amphibian, both of which further improve its ability to live in cool water.

The species has also been found to carry the chytrid fungus, which has devastated frog populations elsewhere. So far, this fungus does not seem to have affected the Lake Titicaca Water Frogs, possibly because the cool temperature of the lake prevents the fungus becoming widely infectious. Should climate change raise the temperature of the lake, then this situation could change and the chytrid fungus may start causing disease and increased mortality.

→ The Titicaca Water Frog is a high-elevation aquatic specialist, perfectly adapted to absorb oxygen from the water through extensive folds of skin which cover its body. It will sometimes perform "push-ups" to force more water through these skin folds and increase oxygen absorption.

LIFE ON LAND AND IN WATER

SCAPHIOPUS COUCHII

Couch's Spadefoot Toad

Desert digger

SCIENTIFIC NAME:	*Scaphiopus couchii*
FAMILY:	Pelobatidae
LENGTH:	2½-3½ in (55–90 mm)
LIFE HISTORY:	Aquatic eggs (3,000 in clumps) and tadpoles; terrestrial adults
NOTABLE FEATURE:	Spadelike structures on the hind feet used to dig a burrow
IUCN RED LIST:	Least Concern

Adapted to desert regions, arid grasslands, savannah, and dry forest areas of southwest North America, Couch's Spadefoot Toad is so named because of the spade-like structures on the hind feet, which are used for digging burrows in order to see out seasonal hot, dry periods. Couch's Spadefoot tends to be a variable green, yellow, or brown, with darker patches or spots. It can be distinguished from other spadefoot toads by the sickle-like shape of the spade on its hind feet.

Spadefoots can remain burrowed and relatively inactive for ten months of the year, until stimulated to emerge by seasonal thunderstorms that herald the onset of the wet season. Following significant rains, within a day or so of emerging, the toads have mated and each female will lay up to 3,000 eggs in the small, rain-filled pools that form in hollows and streambeds. The eggs can hatch in less than a day and tadpoles complete their development in under two weeks. If the tadpoles become too crowded in desiccating pools, they will start eating each other, thereby ensuring that at least some will make it through to metamorphosis.

Likewise, the adults feast on termites and other insects also stimulated by the rains to emerge and reproduce. An adult Spadefoot can consume more than half of its body weight in insects in a single night of foraging, and this may be all it needs to survive for the next ten months. During the long period of inactivity underground, the metabolic rate drops substantially and the toads lower their rate of breathing to further reduce any water loss. The extremely short period of breeding and activity entirely tied to the rains therefore enables the Spadefoot to survive in desert conditions.

→ Couch's Spadefoot Toad is only rarely above ground—perhaps on no more than 20 nights in a given year. During their time on the surface, they feed and breed before retreating back underground, and they can survive exceptionally dry periods by absorbing moisture from the surrounding soil.

LIFE ON LAND AND IN WATER

NEOBATRACHUS AQUILONIUS

Northern Burrowing Frog

Life in suspended animation

SCIENTIFIC NAME:	*Neobatrachus aquilonius*
FAMILY:	Limnodynastidae
LENGTH:	Up to 2½ in (60 mm)
LIFE HISTORY:	Aquatic eggs (up to 1,400 in strings) and tadpoles; terrestrial adults
NOTABLE FEATURE:	At nearly 3 in (8 cm) long, the tadpoles of this frog are longer than the adult
IUCN RED LIST:	Least Concern

The Northern Burrowing Frog is one of many species across several anuran families that has adapted to survive in sparsely vegetated arid and desert environments. These frogs have a rotund body and spade-like structures on the hind feet to aid burrowing, which they do in reverse. They are usually gray to dark brown in color with yellow to orange blotches and mottled markings, and often have a light-colored or yellow stripe running down the back.

Adaptable estivation
The Northern Burrowing Frog can adapt to different drying conditions, and may not produce desiccation-proof layers of shed skin unless conditions dictate (1). If the surrounding soil loses moisture, the frog can detect this and begins to slough its skin, forming a barrier between it and the drying soil (2). In the harshest conditions, it can produce more than 200 layers of shed skin and survive in this watertight cocoon for more than two years (3).

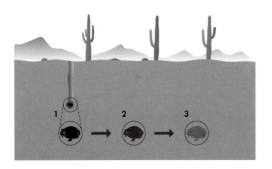

Active nocturnally, and only for brief periods following significant rainfall, Northern Burrowing Frogs emerge from burrows to feed and reproduce, with females depositing eggs that hatch and metamorphose within a matter of weeks. Following these periods of explosive activity, frogs estivate to avoid desiccation by digging themselves vertically into damp sandy or sand/clay soils up to 3 ft (1 m) or more in depth. When the surrounding substrate dries, the frog begins to lay down successive layers of shed skin, enveloping the entire body except for the nostrils and forming an increasingly impermeable cocoon. A single layer of skin can be deposited every two or three days, and frogs can survive in this way for many months, with one study having counted 229 layers—giving an estimated average of 20 months spent dormant underground. When cocooned in this way, the frog's metabolic rate drops to about one-fifth of its normal rate. However, frogs respond to the soil conditions, cocooning only when moisture levels fall below the threshold required to maintain hydration from water stored in their bodies.

→ The Northern Burrowing Frog is known to have four sets of chromosomes (humans have two sets). This additional genetic diversity may mean that it can more readily adapt to climate change.

LIFE ON LAND AND IN WATER

LITHOBATES SYLVATICUS

Wood Frog

Frog that freezes

SCIENTIFIC NAME:	*Lithobates sylvaticus*
FAMILY:	Ranidae
LENGTH:	1½–3½ in (35–85 mm)
LIFE HISTORY:	Aquatic eggs (300–1,500 in clumps) and tadpoles; semi-aquatic adults
NOTABLE FEATURE:	As well as freezing, Wood Frogs can stop their breathing, blood flow, and heartbeat
IUCN RED LIST:	Least Concern

The Wood Frog is perfectly camouflaged against leaf litter on the forest floor, thanks to a highly variable coloration of brown, gray, olive-green, and reddish hues. It is the most widely distributed amphibian in North America, with a vast range spanning damp woodlands of the central and northern United States, including north of the Arctic Circle in Alaska and Canada. It therefore occurs across a wide range of climatic conditions, with Wood Frogs found in the southern part of its distribution tolerating higher temperatures than those in the north.

Within the Arctic Circle the Wood Frog has developed an intriguing physiological mechanism that allows it to freeze solid but stay alive. It does this by using a natural antifreeze to stop ice crystals forming which can damage organs and tissues (as occurs when humans suffer frostbite).

To prevent frostbite and organ malfunction, Wood Frogs produce antifreeze proteins which attach to the ice crystals in the cell spaces and prevent them growing further. In addition, cryoprotectants—glucose and urea (a component of urine)—flow into the cells from the surrounding spaces. The increased concentration of cell solutes means the cellular fluids freeze at a lower temperature, in the same way that seawater freezes at a lower temperature than fresh water.

Wood Frogs in places such as Alaska also have enlarged livers. This is due to the increased quantities of stored glycogen, which can be converted to glucose and used as a cryoprotectant. Likewise, urea is absorbed from the bladder, and rather than being excreted is diverted for use as a cryoprotectant. In this way, Wood Frogs can remain frozen for up to six months in the Alaskan winter at temperatures as low as minus -0.4°F (18°C). Understanding how these frogs use natural antifreeze is helping researchers develop novel ways of preserving and protecting organs for human medical treatments.

→ The Wood Frog is not just the only frog, but the only amphibian known to occur north of the Arctic Circle in the Western Hemisphere.

COMMUNICATION & REPRODUCTION

COMMUNICATION AND REPRODUCTION

Vocal communication

→ The Ankafana Bright-eyed Frog (*Boophis luteus*) from Madagascar has paired vocal sacs, and makes a tuneful whistling vocalization that can last for several minutes.

↙ During the breeding season, male Indus Valley Bullfrogs (*Hoplobatrachus tigerinus*) take on a distinctive yellow coloration, which contrasts strikingly with their blue vocal sacs.

Frogs were the first animals with vocal cords but are the only amphibians with an extensive repertoire of vocal communication. Although mainly used by males during reproductive periods, some females also vocalize, and many frogs employ specific calls in defense of their territories or in response to predators.

Irrespective of the species, the reproductive vocalizations—advertisement calls—of anurans are one of the most well-known animal sounds on the planet, but why do they generate these sounds, what are they saying, and how are they doing it? Fundamentally, frog mating calls are mechanisms for reproductive isolation—that is, they appeal only to female frogs of the same species. Accordingly, frog vocalizations are important tools in anuran taxonomy and often the only means of differentiating between species. This is especially the case when morphologically similar species live in sympatry, as do several of the North American *Dryophytes* treefrogs, for example.

Even within species there can be significant variation in advertisement calls, and this is most often present in frogs with a wide geographic range where populations may be subject to different environmental conditions (North America's Northern Cricket Frogs, *Acris crepitans*, for example). However, regional/geographic differences are usually reflected in quantifiable traits not directly related to call structure, such as the frequency (measured in hertz), repetition rates, or note duration. At least one family of frogs, New Zealand's Leiopelmatidae, do not make advertisement

VOCAL COMMUNICATION

calls at all, instead using chemical cues to communicate. This mode of communication may be an addition to the repertoire employed by many anurans, and is perhaps more widespread than currently recognized.

If not delivering an aggressive or defensive call, most frogs vocalize without opening their mouths, although some—like the giant or slippery frogs (genus *Conraua*), from West Africa, and the Southeast Asian splash frogs (genus *Staurois*)—do call with the mouth open. Although calls are emitted as part of the respiratory cycle, vocalizing is an energetically costly activity for frogs and often requires considerable effort. Notable exceptions to this are the frogs in the fully aquatic Pipidae family, which do not possess vocal cords, instead using their modified larynx as an internal vocal sac.

The process of vocalizing in frogs

The main structures used are the nostrils, buccal cavity, lung, larynx, and vocal sac. The pulmonary respiration cycle (frogs also breathe through their skin) is powered by expansion and contraction of muscles in the floor of the buccal cavity (known as the buccal pump) as air is forced into and out of the lungs, aided by corresponding opening and closing of the nostrils. When vocalizing, this cyclic system is closed, and a column of air is instead driven alternately between the lungs and vocal apparatus forcing open the larynx and vibrating the vocal cords.

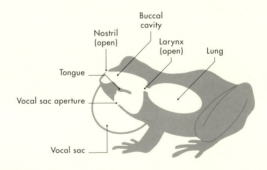

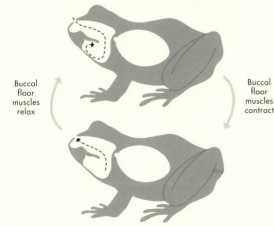

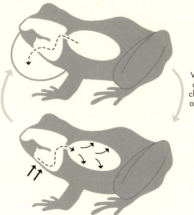

COMMUNICATION AND REPRODUCTION

↑ The Hourglass Treefrog (*Dendropsophus ebraccatus*) from Mexico produces a variety of calls to advertise its presence to females, and to intimidate rival males.

VOCAL COMMUNICATION

↑ Many frogs (like this European green frog, *Pelophylax* sp.) often call while in the water, inflating their vocal sacs to advertise their presence to other frogs.

COMMUNICATION AND REPRODUCTION

TYPES OF CALL

There are two modes of anuran advertisement calls: simple calls and complex calls. A simple call is one that consists of a single (often repeated) note, while a complex call is one comprised of multiple different notes. Some frogs vocalize only simple calls, others combine a simple call with a suite of more complex notes, and others emit only complex calls. The Madagascar Bright-eyed Frog (*Boophis madagascariensis*) has a repertoire of 28 different note types, rivaling (and indeed exceeding) that of many birds. Indeed, many anuran vocalizations are comparable to that of birds, and cover a range of sounds, including barks, clicks, quacks, trills, whistles, and even rasping booms that carry for several miles. Many frogs have common names that represent the sound of their advertisement call, such as the Motorbike Frog (*Litoria moorei*), Eastern Banjo Frog (*Limnodynastes dumerilii*), Carpenter Frog (*Lithobates virgatipes*), and Hooting Frog (*Heleioporus barycragus*).

↑ Top, the Madagascar Bright-eyed Frog (*Boophis madagascariensis*) has an extensive and complex vocal repertoire, with 28 different note types recorded to date. Bottom, the Western Green and Golden Bell Frog (*Litoria moorei*) from Australia has a distinct vocalization that audibly demonstrates its other common name, the Motorbike Frog.

Vocalizations

A) The simple, single-note call of a sooglossid frog, in this case Gardiner's Seychelles Frog (*Sechellophryne gardineri*) lasting approximately 0.17 seconds; B) the simple multi-note vocalization of the (unrelated) Seychelles Treefrog (*Tachycnemis seychellensis*) lasting around 2.34 seconds; C) the complex multi-note call of another sooglossid, the Seychelles Frog (*Sooglossus sechellensis*) comprised of a pulsed primary note followed by a series of different secondary notes (8 in this case) giving a call length of 1.24 seconds; D) the pulsed single-note call of *Sooglossus sechellensis* (so no secondary notes), lasting 0.27 seconds.

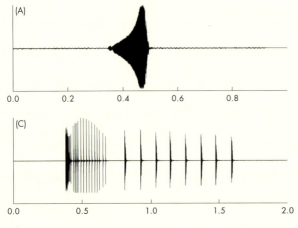
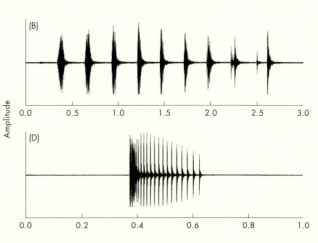

VOCAL COMMUNICATION

HOW FROGS HEAR

A section of the middle ear forms the outermost part of anuran hearing anatomy, visible in many frogs as the circular tympanic membrane—the ear drum—on either side of the head behind the eye. The tympani (singular: tympanum) channel vibrations (sound) through the middle ear to the inner ear for transmission to the auditory centers of the brain.

However, this is not the only sound-transmission pathway for frogs. Known as extratympanic pathways, the body wall adjacent to the lungs can transmit sound vibrations to auditory pathways in the head, as can bone conduction via the oral cavity, forearms, and shoulders. In addition, there is a further pathway unique to amphibians, the opercularis system, which may also aid detection of low-frequency sound transmission via the substrate. Extratympanic pathways are crucially important for those frogs that are effectively earless—lacking external ears (for example, several microhylids and leptodactylids) and sometimes even middle ear structures (several bufonids, including members of the genus *Atelopus*—see the Panamanian Golden Frog, *Atelopus zeteki*, page 98—and all the sooglossids).

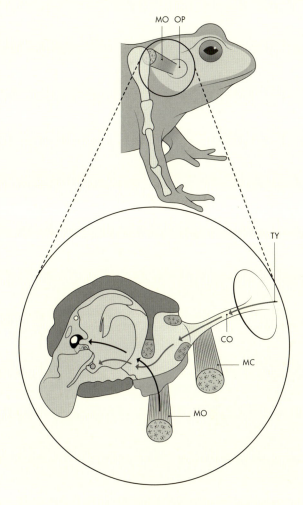

The peripheral sound transmission pathway in frogs

A cross-section schematic of the right inner-ear of a typical anuran, with light gray arrows indicating the primary pathway of sound transmission from the tympanum (TY) through to the inner ear via the columella (C), aided by contractions of the columella muscle (MC). The opercularis muscle (MO) connects to the operculum (OP), but originates on the shoulder skeleton, forming a secondary auditory pathway (black arrows) known as the opercularis system.

← In many frogs, like this American Bullfrog (*Lithobates catesbeianus*, see page 96), the external eardrum, or tympanum, is much larger in males than females.

COMMUNICATION AND REPRODUCTION

Visual communication

Although nearly all frogs are highly vocal animals, in some circumstances elaborate vocal communication may not be enough to intimidate rivals, attract mates, and defend territories. Consequently, visual signaling has developed in at least seven different families of frogs in different parts of the world.

← As well as generating sounds that signal to rival males and receptive females, frogs have other ways of communicating. In hyperoliid treefrogs such as the Cinnamon-bellied Reed Frog (*Hyperolius cinnamomeoventris*), a gland on the vocal sac produces scent, which is thought to be emitted while the frog is calling.

Visual communication can take many forms and frogs have evolved several different methods to signal to females and defend their territories. For example, Krefft's River Frog (*Phrynobatrachus krefftii*) lives along East African mountain streams and is well-camouflaged against the rocks and vegetation. When breeding, to make themselves stand out from their camouflage and be seen by other frogs, males will inflate their bright yellow vocal sac, often doing this without vocalizing.

Other frogs have adopted a range of different body postures to signal their intent. Probably the most remarkable visual signaling occurs in frogs that use their arms, legs, feet, and sometimes all three to communicate. Hind legs are often used and raised to show the feet with toes stretched apart, revealing a flash of bright color in the webbing between them. The Sabah Splash Frog (*Staurois latopalmatus*) and the Bornean Black-spotted Rock Frog (*S. guttatus*) both extend and rotate alternate legs (sometimes both!) to flag their bright white/blue-white foot webbing in territorial defense and to attract a mate.

VISUAL COMMUNICATION

In other species—for example, the Panamanian Golden Frog (*Atelopus zeteki*, see page 98)—both males and females will use semaphore comprised of arm and leg raises to communicate. If visual signaling does not ward off a rival, male frogs will often wrestle with their competitors, and some species have even evolved weapons to battle with other frogs.

The prevalence of such visual signaling seems to depend on the environment. Visual communication may provide more precise information about the location of an individual than vocal communication in some species. Moreover, in those frogs that breed along fast-flowing streams it may be difficult to be heard above the noise of rushing water. Either way, no frogs that use visual communication have completely abandoned vocalization, so the relative importance of the two modes of communication depends on whether it is easier to be seen or heard in a particular habitat.

↑   The Kottigehar Dancing Frog (*Micrixalus kottigeharensis*) from the Western Ghats of India is so-called because it rhythmically extends its hind legs to flash its foot-webbing. Called foot-flagging, it does this to attract female frogs, and to deter competing males.

FROG WEAPONRY

The wrestling matches between male frogs rarely result in serious injury as they mainly involve pushing and kicking. If an existing male is already grasping a female, another male may try and wedge himself between the pair and lever the first male off. Likewise, frogs do not usually aggressively bite each other as they lack the teeth or jaw structure to inflict a wound, although the male African Bullfrog (*Pyxicephalus adspersus*, see page 130) is one of many species which have tooth-like projections (odontoids) on the lower jaw. These "fangs" are not true teeth, but grow as protuberances from the jawbone, and are used in duels with other males and in some cases as defense. Frogs in one genus, *Limnonectes*, are even known as the fanged frogs, although this feature is not present in all species.

Several frogs have taken fighting to another level and developed weapons. Some leptodactylid frogs, for example, have sharp spines on their front feet which they use to jab and lacerate their opponents. Others are equipped with spines that project from the forearms, as seen in many of the charismatic glassfrogs (see the Nicaragua Giant Glassfrog, *Espadarana prosoblepon*, page 100). In one genus of hylid treefrogs (*Boana*)—commonly referred to as gladiator frogs—male frogs fight over breeding sites, using their thumb spikes in wrestling matches to stab a challenger in the body and head, on occasion killing their opponent.

→ Most frogs show little aggression directed at other species. However, in defense of its territory and offspring the African Bullfrog (*Pyxicephalus adspersus*) will pursue and attack other frogs and potential predators, including humans.

VISUAL COMMUNICATION

COMMUNICATION AND REPRODUCTION

Selection, satellites, and sneaky sex

Charles Darwin coined the idea of "sexual selection," whereby some males have characteristics that make them more successful in achieving matings than others, and so pass on those characteristics to the next generation. Differences in male breeding success are determined either by males securing a mate to the detriment of another male or by females selecting males with certain characters, both of which are modes employed by frogs.

SELECTION, SATELLITES, AND SNEAKY SEX

→ A female Gray Foam-nest Treefrog (*Chiromantis xerampelina*) of southeastern Africa makes a foam nest on leaves, and four smaller attendant males compete to fertilize the eggs.

↙ In many frog species, including American Toads (*Anaxyrus americanus*), competition for mates is often keen, and sometimes mating balls are formed comprising of several males trying to secure amplexus with a single female.

For sexual selection to work there must be sufficient variation in the calls and displays for males to assess their chance of outcompeting other males, and for females to make an appropriate choice of mate. This requires effective messaging both within and between the sexes. Moreover, the types of calls and displays need to be a good indicator of the fitness of the frog that is calling and displaying. Far from being restricted in this mode of communication, there is enormous variation in the vocalizations used by frogs. In contrast, other amphibians—the salamanders and caecilians—are practically silent.

↑ In Costa Rica, several hundred Misfit Leaf Frogs (*Agalychnis saltator*) can converge on communal spawning sites, attaching their clutches of eggs to leaves overhanging temporary pools that form at the start of the rainy season.

However, frogs do not always rely on vocalizations to convey information about themselves to other frogs, and some would-be suitors often take a chance on benefiting from the actions of others in order to secure a mate, a behavior sometimes referred to as sneaky mating, or sexual parasitism. For example, male American Green Treefrogs (*Dryophytes cinereus*, see page 64) will position themselves as a non-vocalizing satellite to a vocally active frog, with the aim of intercepting a female attracted by the other male's calling efforts. Male frogs will also switch between satellite and calling strategies, often during the same night. Through their vocalizations, sometimes combined with visual displays, males often attract females while simultaneously intimidating rival suitors. But these roles are sometimes reversed, and it is the female frogs that compete for males, and often also make reproductive vocalizations, as seen in the Mallorcan Midwife Toad (*Alytes muletensis*, see page 102).

→ In some populations of European Common Frogs (*Rana temporaria*) a female's eggs may not necessarily be fertilized by the amplectant male.

SELECTION, SATELLITES, AND SNEAKY SEX

MULTIPLE PATERNITY

Having secured his mate in amplexus, the male sheds his sperm over the eggs as they are laid by the female. With all the effort he has put into calling, seeing off other males, and finding his mate, the male's reward is that the offspring should all be his. However, this may not be the case. In explosive breeders (see pages 92–3) that use communal spawning sites, all the wrestling may mean unpaired males can sneakily release their sperm over some of the eggs. In one Pyrenean population of European Common Frogs (*Rana temporaria*, see page 32), it was found that some males actively try and displace egg masses as they are released by a female in amplexus with another male, so they can shed their own sperm over them, or even crawl into a recently laid clump of spawn to access potentially unfertilized eggs. Such behavior is known as "clutch piracy."

Likewise, in some foam-nesting frogs, which churn secretions and sperm into a frothy mass in which the eggs are laid, some sneaky males may also engage in the churning and manage to get their sperm into the egg mass. In some frogs, 20–50 percent of the offspring in a clutch may be fertilized by more than one male. Such multiple paternity may increase the genetic diversity of the resulting tadpoles.

COMMUNICATION AND REPRODUCTION

Explosive and prolonged breeding

Some frogs have short, sharp breeding periods where mating and egg laying are over in just a few days of frantic activity. In others, the breeding period is longer and more leisurely, taking place over a period of several weeks. Which strategy is used depends on the prevailing climate and how much competition there is for mates.

Tales of "raining frogs" date as far back as biblical times. Some of these accounts are reliable and describe tornados sucking up water from a pond or lake and then depositing the accompanying residents (fish or frogs) elsewhere. Such meteorological events are rare but have led to the sudden appearance of frogs after rainfall being attributed to them falling from the clouds.

However, there is a more biological explanation for a sudden abundance of frogs after rain. Rainfall will stimulate frogs to emerge from hibernation or estivation—often in substantial numbers—and drive them to converge on their breeding pools. Indeed, once ponds and streams fill after the spring rains, it may be a race against time for frogs to find a mate, lay eggs, and provide the opportunity for their tadpoles to successfully complete development before the water body dries up.

→ European Common Frogs are typical explosive breeders, and many hundreds of frogs may simultaneously converge on a pond and complete spawning within a matter of days.

EXPLOSIVE AND PROLONGED BREEDING

COMMUNICATION AND REPRODUCTION

With so many frogs converging on a breeding pool at the same time, such "explosive breeding" may be completed very quickly. In contrast, in more permanent water bodies the breeding period may last longer, with different individuals arriving and departing at different times.

In general, whether frogs are "explosive" or "prolonged" breeders depends on the species, but there is some variation in this pattern. The European Common Toad (*Bufo bufo*, see page 104), for example, breeds at different times in different parts of its wide range, but when it does so it is always an explosive breeder with all mating and egg laying usually over in 10–14 days. However, the Natterjack Toad (*Epidalea calamita*, see page 246) has a prolonged breeding period in the cooler northern European parts of its range, but where breeding ponds hold water for a shorter period in the warmer south, it becomes an explosive breeder.

EXPLOSIVE BREEDING

In explosive breeders, the females arrive, secure a mate, lay eggs, and then leave, while the males will hang around for longer in the hope of securing another partner. This means that males outnumber females at the breeding site, which results in a lot of competition (aptly referred to as scramble competition) between males for a female, and elaborate wrestling matches ensue between the males in an attempt to get a secure grip on a female (see page 94). Consequently, there is not much opportunity for the females to be choosy in explosive breeding species. Indeed, a female can

EXPLOSIVE AND PROLONGED BREEDING

↖ A male Common Midwife Toad (*Alytes obstetricans*) wrapping the eggs around his back legs as they are produced by the female.

↑ A pair of Borneo Narrow-mouthed Frogs (*Microhyla nepenthicola*) may lay multiple clutches of eggs over many days within a single pitcher plant.

↗ The American Bullfrog (*Lithobates catesbeianus*) can lay up to 20,000 eggs in a single clutch.

sometimes drown as a result of several over-amorous males all struggling to get into the mating embrace with her at the same time.

PROLONGED BREEDING

In frogs with more prolonged breeding periods there is still competition between males, but also more opportunities for females to choose a mate. Males usually call in choruses, with the call of one or more individuals triggering others (in fact, other noises such as a plane flying overhead, a dog barking, or even the human voice can sometimes stimulate a group of frogs to burst into song). The collective chorus of the males provides a location for migrating females to aim for. Within the choruses the males are usually spaced out.

EMBRACE YOUR PARTNER: TYPES OF AMPLEXUS

To ensure a successful mating, the male frog must hang on to his chosen mate until she is ready to lay eggs. He then sheds his sperm over the eggs (this is external fertilization). It is thought that early on in the evolution of frogs, males held the female around the waist in a position known as inguinal amplexus. This position is still employed by a variety of frog taxa, including the "ancient" families Leiopelmatidae and Ascaphidae.

The development of axillary amplexus, where the male holds the female around her front legs with his forefeet tucked into her armpits, is thought to have evolved later. Indeed, males develop rough pads on their thumbs to help secure their grip. It is difficult for an amplectant male to be dislodged by another male (or by a curious human), and such is the tightness of the grip that when the pair separate "scars" may be visible in the female's armpits. Although inguinal and axillary amplexus are the predominant forms of mating embrace, there are a variety of other ways of maintaining contact within pairs. Some rocket frogs (genus *Colostethus*) and dart frogs (genus *Dendrobates*), for example, use cephalic amplexus, where the male's hands are pressed around the female's neck. In the microhylid rain frogs of southern Africa (genus *Breviceps*), the male is so much smaller than the female that his arms are too short to reach around her rotund body, but a glue-like skin secretion helps to keep his belly stuck to her back.

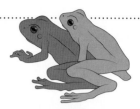
Inguinal

Head straddle

Loose amplexus

Axillary

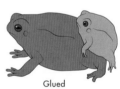
Glued

Cephalic

Mating positions used by frogs

At least nine different amplectic positions are used by modern anurans. Inguinal amplexus—found in archaeobatrachian frogs—is considered the ancestral position, although it is also present in several neobatrachian lineages. Axillary amplexus is thought to have evolved from inguinal amplexus, and is the position found most commonly in living frogs.

Gular

Dorsal straddle

No amplexus

Some species use a "lek" system, whereby males call from and defend a space they have selected as being a good point from which to call. Many treefrogs form choruses by calling from elevated perches in trees and bushes. Such positions help transmit the calls over greater distances.

Once a female treefrog has selected a male to mate with, she carries her suitor down to a suitable egg-laying site. So, although the lek system means that males may be spaced out within the chorus, there may be a communal spawning area that provides optimal conditions for the eggs of all females to develop. In other species, the spaces that are defended are true territories, in that they may comprise particularly good areas for females to lay their eggs. In such cases, the best egg-laying sites will be defended by the dominant (usually larger) males (see the American Bullfrog, *Lithobates catesbeianus*, page 96, for example).

↑ A male and female White-eared Treefrog (*Feihyla kajau*) in axillary amplexus.

LITHOBATES CATESBEIANUS

American Bullfrog

The largest frog in North America

SCIENTIFIC NAME:	*Lithobates catesbeianus*
FAMILY:	Ranidae
LENGTH:	4½–7½ in (110–190 mm)
LIFE HISTORY:	Aquatic eggs (up to 20,000 in clumps) and tadpoles; semi-aquatic adults
NOTABLE FEATURE:	Taste receptors that change with the frogs' diet, from the tadpoles (largely herbivorous) to the adults (carnivorous)
IUCN RED LIST:	Least Concern

The American Bullfrog is widely distributed across North America, where it occurs in a variety of water bodies, although it prefers warmer habitats that don't dry out in summer. It is a large, green frog, with a general coloration that can vary according to geographic location with yellowish, brown, or gray patterning. A voracious and opportunistic predator, the American Bullfrog consumes a wide range of prey, ranging from insects to small mammals and other frogs, including individuals of its own species. As such, it is often the dominant frog species within a wetland system.

The males are territorial and will vocalize in a well-spaced-out chorus. Their calls are particularly loud, and larger males with the best territories—warmer patches of water with dense aquatic vegetation, ideal for egg deposition—will aggressively defend them from other males. However, within a territory, male frogs recognize the calls of their neighbors and respond more vigorously to the calls of strangers, meaning they don't waste energy interacting with frogs that already have an adjacent territory. The chorusing of male frogs attracts females to the breeding site, and there they select a male which holds the best territory for depositing eggs and development of young. Females become more discerning as they get older, selecting the largest males that hold the best territory.

The mating system of the American Bullfrog differs from that of many explosive breeders (where scramble competition leads to a free-for-all between the males trying to mate with females, with no territory or valuable breeding area at stake). However, in bullfrog choruses, the large, dominant, territory-defending males don't always have it their own way. Around the periphery of male territories, smaller, subordinate males wait. These satellite males often do not vocalize and can therefore go undetected by the territory-holding males. The aim of satellite males is to intercept females before they get to a territory-holding male. Although females may try and shake off the attention of the satellites, some of these smaller males are successful in securing a mating. A further advantage of being a satellite is that if the territory is vacated by the dominant male (he could get eaten by a predator), the satellite has an opportunity to take over that territory. Although, this may be temporary, as he may subsequently be usurped by a larger male!

→ The largest frog in North America, the American Bullfrog has an unmistakably loud, low-frequency advertisement call. It has been introduced across the globe (including to western North America) for entertainment (such as frog-jumping contests), human consumption, and pest control.

COMMUNICATION AND REPRODUCTION

ATELOPUS ZETEKI

Panamanian Golden Frog

A national symbol and cultural icon in Panama

SCIENTIFIC NAME:	*Atelopus zeteki*
FAMILY:	Bufonidae
LENGTH:	1½–2¼ in (35–55 mm)
LIFE HISTORY:	Aquatic eggs (up to around 600 in a single string) and tadpoles; terrestrial adults
NOTABLE FEATURE:	The skin of a single Panamanian Golden Frog contains enough toxin to kill 1,200 mice
IUCN RED LIST:	Critically Endangered

One of the harlequin frogs (genus *Atelopus*), this striking yellow frog is often patterned with black markings and known from rocky torrents in the upland rainforests of central Panama. Younger frogs are well-camouflaged against the moss-covered rocks, and it is not until adulthood that they adopt their remarkable coloration—an example of aposematism: advertising that these frogs are toxic to would-be predators. The Panamanian Golden Frog is considered the most toxic of the harlequin frogs.

Although the Panamanian Golden Frog vocalizes, it is one of the "earless" frogs, as it lacks external and middle-ear structures (see page 81). The loss of these structures, and their diurnal behavior, may have evolved alongside their toxicity. Male frogs adopt territories along exposed rocky boundaries of streams and waterfalls—where they would otherwise be vulnerable to predation—but because of the noise of rushing water cannot rely on acoustic communication alone. In fact, they prefer to use semaphore signaling, incorporating a sophisticated series of hand-waving and foot-raising with which to deter rival males and attract females.

Intriguingly, female frogs also sometimes participate in these displays, possibly to attract the attention of males, or to rate or test the semaphore skills of a male. The territoriality of male frogs can be such that they will semaphore toward non-frog sources of disturbance, including humans. Shortly after this unique behavior was first captured on film for David Attenborough's *Life in Cold Blood* series in 2006, the species succumbed to the chytrid fungus and was wiped out in the wild (see Chapter 9). Fortunately, some frogs were collected for a conservation breeding project just in time and the species currently survives in captivity.

→ The Panamanian Golden Frog is a national symbol in Panama. Every year on August 14, National Golden Frog Day is celebrated around the country.

COMMUNICATION AND REPRODUCTION

ESPADARANA PROSOBLEPON
Nicaragua Giant Glassfrog
See-through skin

SCIENTIFIC NAME:	*Espadarana prosoblepon*
FAMILY:	Centrolenidae
LENGTH:	¾–1¼ in (20–30 mm)
LIFE HISTORY:	Terrestrial eggs (up to 20 in a single layer) and adults; aquatic tadpoles
NOTABLE FEATURE:	Male frogs have long spines for use in combat
IUCN RED LIST:	Least Concern

With their large prominent eyes, translucent skin, and general green coloration, glassfrogs are unmistakable, and often aligned with that most famous of frogs, Kermit. The Nicaragua Giant Glassfrog is emerald-green in color, with dark scattered flecks, but can have yellow dots, an unmarked back, or dark rings with yellow centers on a green back. It is found in densely vegetated tropical montane forest areas in riparian zones from Central America to northwestern parts of South America.

Male Nicaragua Giant Glassfrogs are arboreal and territorial, and distribute themselves at intervals along their riparian habitat by vocalizing. If a male enters the territory of another, and the latter remains undeterred, a wrestling match can ensue. Such battles can last for up to 30 minutes, only ending when one male either submits by flattening himself against a leaf or drops from the vegetation from which both contestants have been suspended. During these matches, the males use large, blue spines connected to the humeral bones (humeral spines) and attempt to push off their rival. The spines of each male appear to interlock in combat, and their purpose may be primarily for leverage. However, injury from humeral spines has been reported in other glassfrogs, so their use may be for both purposes. Interestingly, females have been reported to present an underdeveloped humeral spine.

Like many treefrogs, the Nicaragua Giant Glassfrog lays its eggs on leaves, vegetated rocks, or branches which overhang water. When the eggs hatch, they drop into the water below to feed, grow, and develop.

→ Like other glassfrogs, the Nicaragua Giant Glassfrog has translucent skin, in particular on the underside. This is especially important when at rest on leaves, as their outline could easily be seen by a predator below. When the frogs rest in this way, their red blood cells retreat into the liver, making them almost invisible.

COMMUNICATION AND REPRODUCTION

ALYTES MULETENSIS

Mallorcan Midwife Toad

One of the few Critically Endangered frogs to have been downlisted, thanks to conservation efforts

SCIENTIFIC NAME:	Alytes muletensis
FAMILY:	Alytidae
LENGTH	1¼–1½ in (30–40 mm)
LIFE HISTORY	7–12 eggs wound around the back legs of the male; aquatic tadpoles; terrestrial adults
NOTABLE FEATURE	The females compete with each other for access to males
IUCN RED LIST	Endangered

The endemic Mallorcan Midwife Toad is a relatively small archaeobatrachian frog, with a greenish-gray background coloration, patterned with darker green or brown patches. It was considered extinct for nearly 6,000 years. Then, in 1979, some tadpoles were found in water bodies hidden in remote mountain gorges in the north of the island. Given the precarious state of the species at this stage, some adult toads were collected for a captive breeding project, allowing for important research on reproduction, genetics, and responses to threats, as well as generating a large number of toads for release back into the wild.

Including *Alytes muletensis*—which is only found on Mallorca—the midwife toads comprise a group of six species, with four in western Europe (*A. obstetricans, A. cisternasii, A. dickhilleni,* and *A. almogavarii*) and one species in north Africa (*A. maurus*). Courtship behavior involves both sexes of Mallorcan Midwife Toads vocalizing to advertise their interest in finding a mate. In fact, only females who can hear calling males will continue to mature their eggs ready for mating.

Moreover, females will approach calling males and choose a male based on the attractiveness of his call (low-frequency calls tend to be made by larger, fitter males in many species of anurans). Thus, it is the females who actively control the progress of courtship, and they will also wrestle with each other over a desired male. This role-reversal extends to parental care of the eggs, which is practiced by the male in all six species. After the male has fertilized a string of eggs deposited by the female, her task is complete. The male then winds the eggs around his back legs and cares for them until they are ready to hatch.

This role reversal between males and females—with females fighting over males and males caring for the eggs—seems to be related to the unusual mating system. Unlike many other anurans, the sex ratio of the midwife toads is biased toward females and the availability of males to carry the eggs is the limiting factor for reproductive success. In such cases, the males are worth fighting over.

→ Like many frogs, the Mallorcan Midwife Toad is threatened by the disease-causing fungus *Batrachochytrium dendrobatidis* (*Bd*). In a world first, the fungus was eradicated from the wild habitats of this frog using a common fungicide (itraconazole).

COMMUNICATION AND REPRODUCTION

BUFO BUFO

European Common Toad

Rafting toads have repeatedly colonized Scottish islands

SCIENTIFIC NAME:	*Bufo bufo*
FAMILY:	Bufonidae
LENGTH:	2–4 in (50–100 mm)
LIFE HISTORY:	Aquatic eggs (up to 5,000 in strings) and tadpoles; terrestrial adults
NOTABLE FEATURE:	Toxic tadpoles mean these toads can breed in ponds with fish
IUCN RED LIST:	Least Concern

Found across most of Europe, parts of North Africa, and ranging well into eastern parts of Russia, the European Common Toad is a familiar anuran to many, often being present in parks, gardens, farmland, grasslands, and forested areas. Very much a "typical" toad (Bufonidae are the "true" toads), with skin covered in numerous tubercles, it has striking, copper-colored eyes and a body color ranging from gray to brown, olive, and reddish, often with darker spots.

Like many frogs in temperate zones, the European Common Toad is an explosive breeder, with males and females migrating to their breeding ponds in the spring for an intense period of reproduction and spawning that is completed in just a few days. Males tend to arrive first to increase their chances of intercepting a female on her way to the breeding site. So amorous are the males, they will often mistakenly grab other male toads. However, those on the receiving end will vocalize a release call, which informs the over-amorous suitor that he has got it wrong!

For many male toads then, finding a female can be a process of trial and error, with success only verified if the toad he grabs remains silent. Coincidentally, after spawning, females will also produce a release call to discourage other male toads. Although males do call to attract females, much of the vocal activity heard at breeding sites stems from male release calls. Indeed, during the breeding period, gently stroking a male European Common Toad on its back is all that is needed for him to announce his gender.

Given the general male-bias at breeding sites, there is often heavy competition for females, which can result in "mating balls" where multiple male suitors attempt to breed with a single female. Larger males tend to be better at dislodging smaller ones and are therefore more successful in some populations. However, size can be a disadvantage, especially when the female is smaller, as the male may not be in the correct position to fertilize her eggs as they are released. Therefore, in some European Common Toad populations there is tendency toward "assortative mating" (where the size of reproducing males and females correspond), while in others there is no relationship between the size of male and female toads, and mating is more random.

→ Although widespread across Europe, the Common Toad has been declining locally, which has largely been driven by loss of habitat and the drying of wetlands.

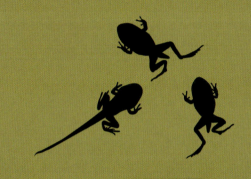

EGGS, TADPOLES, & PARENTING

EGGS, TADPOLES, AND PARENTING

Egg laying, nests, and nurseries

Laying eggs in water, which then hatch into free-living tadpoles that swim and feed independently, is a trait thought to be present in the early evolutionary history of frogs. Today there are frogs that lay eggs on land as well as in water; frogs that feed and care for their tadpoles; and frogs that give birth to fully developed froglets. These modes of development are driven by the availability of water and the prevailing climate, the risks of predation and competition, and food availability for developing young.

LAYING EGGS IN WATER

Most frog species lay their eggs in ponds or ditches that fill following seasonal rainfall. Egg development depends on environmental conditions, particularly temperature, with the eggs hatching into tadpoles that feed independently and metamorphose into tiny froglets within a few weeks. By this time, the pond or ditch may have started to dry up. However, this classical pattern of frog development has evolved in a range of intriguing directions, resulting in some species that don't breed in water at all and have no free-living tadpole stage in their life cycle.

← In the Mallorcan Midwife Toad (*Alytes muletensis*), the male winds the eggs around his back legs and cares for them until they are ready to hatch.

→ The Agile Frog (*Rana dalmatina*) spawns in large individual clumps often attached to a twig or plant stem.

EGG LAYING, NESTS, AND NURSERIES

EGGS, TADPOLES, AND PARENTING

Depending on the species and habitat, eggs are laid in large clumps, long strings, singly, or in small groups. Unlike reptiles and birds, frogs do not lay waterproof, hard-shelled eggs. Instead, the developing embryo floats within a jelly-like capsule that may absorb water and swell after deposition. That said, those frogs that lay eggs on land provide them with a more robust outer capsule than frogs that lay eggs in water. Frogs that lay eggs in warm, sunny places, such as shallow water, have eggs darkly pigmented with melanin, which provides some protection against ultraviolet radiation and may assist with heat absorption. On the other hand, species that lay their eggs in concealed places, such as under stones, in vegetation, or wrapped in leaves, have no need for such protection and hence lack pigmentation.

↖↑ Frogs have a range of methods of hiding and protecting their eggs until they hatch. Some phyllomedusid treefrogs will lay eggs on a leaf, and then fold the leaf around the clutch for protection (left). Glassfrogs such as the Grainy Cochran Frog (*Cochranella granulosa*) will use vegetation in a different way, attaching eggs such that they form a cascading drip under the leaf, thereby protecting them from above, while also allowing rainfall and moisture to flow over the egg mass until the hatching tadpoles drop into the water below (right).

EGG LAYING, NESTS, AND NURSERIES

OVIPARITY, OVOVIVIPARITY, AND VIVIPARITY

Most frogs lay eggs in which the developing embryo is initially nourished by a supply of yolk before it hatches into a tadpole. Such egg laying is termed oviparity. In contrast, rain frogs (the largely neotropical family Eleutherodactylidae) are direct developers, which means the eggs are laid in damp places on land, then development and metamorphosis takes place entirely within the egg and the offspring emerge as fully transformed froglets. The absence of a free-living tadpole phase has enabled such frogs to colonize and thrive in places where there may be little or no free-standing water. However, one species (now sadly considered extinct) went one stage further. The female Puerto Rican Golden Frog (*Eleutherodactylus jasperi*) retained the eggs within her reproductive tract and gave birth to up to five froglets. Such egg retention where the young live entirely on the yolk is termed ovoviviparity. This pattern of giving birth to fully developed young froglets is also found in a few species of toads in the genus *Nectophrynoides* from East Africa.

Once also included in the genus *Nectophrynoides*, the Western Nimba Toad (*Nimbaphrynoides occidentalis*) has added an extra facet to this strategy. Like other ovoviviparous frogs, the eggs are retained in the female oviduct, but once the yolk supplies have been used up the developing tadpoles start feeding on secretions from the lining of the mother's uterus. As the mother is feeding her developing offspring, this type of development is true viviparity, similar to that observed in placental mammals. This mode of development is related to the highly seasonal environment in which the Western Nimba Toad lives. After mating, the female retires underground in the dry season and remains inactive for several months while her offspring gestate. The young are born when the rains return, which coincides with a flush of invertebrate prey for them to feed on.

The different egg masses of frogs
Frogs have evolved multiple modes of egg deposition, from individual eggs and small hidden clusters to communal laying of large clumps or strings, and foam nests on vegetation, at pond edges, and hidden in underground chambers.

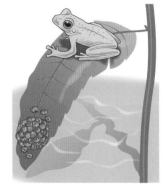
Small clumps on vegetation overhanging water

Large clumps in water

Strings of eggs laid in water

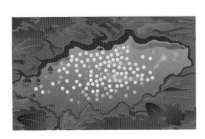
Eggs in an aquatic chamber

Arboreal foam nest

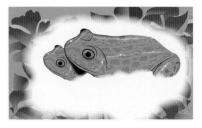
Terrestrial foam nest

LAYING EGGS ON LAND

Frogs that lay their eggs on land tend to lay fewer and larger eggs than those that lay in water. This is because the developing tadpoles must spend longer within the egg capsule and be nourished by a larger supply of yolk than those that can hatch early into water and start feeding independently. Consequently, direct development—the process by which tadpoles complete their development and metamorphose into froglets within the egg capsule—is common in those frogs that lay eggs on land.

↖ The male Bombay Night Frog (*Nyctibatrachus humayuni*) calls from a leaf perch to attract multiple females, and he may end up collecting and guarding many clutches of eggs.

↑ To look after her young, the female Giant Marsupial Frog (*Gastrotheca ovifera*) has a brood pouch on her back with left and right chambers. After fertilized eggs have been pushed into the pouch by the male, they become fully enclosed, and the pouch is subsequently thought to function in a similar way to a mammalian placenta.

↗ A female Harlequin Tree Frog (*Rhacophorus pardalis*) with two males, both of whom are competing to fertilize the eggs contained within the frothy mass.

EGG LAYING, NESTS, AND NURSERIES

FOAM NESTS AND BACKPACK NURSERIES

Foam nests are constructed by frogs from several different families and can be made on the surface of water, among aquatic plants, on land, or in trees close to water. As the eggs are laid and fertilized, the male or female, or sometimes both, whips them into a foamy mass with the aid of secretions and by a paddling action of the legs. The outer layers of foam then dry to form a rubber-like or crusty protective coat. Inside the nest, the foam protects the developing embryos against drying out and extremes of temperature and may even provide some protection against infection by pathogens. In two groups of frogs, the eggs are transferred to the back of the female who carries them around while they develop. In the marsupial frogs of Central and South America (genus *Gastrotheca*), the male pushes the eggs into a pouch on the back of his mate. These hatch in the pouch and, depending on the species, later emerge either as well-developed tadpoles or completely metamorphosed froglets. In the Surinam toads (genus *Pipa*) each egg is absorbed into an individual fleshy pocket on the back of the female where they complete their development (see the Surinam Toad, *Pipa pipa*, page 126).

EGGS, TADPOLES, AND PARENTING

EGG CLUTCH SIZE

The majority of frog species do not care for their eggs or tadpoles, but to ensure the survival of at least some young, one strategy is to lay enormous clutches of thousands of eggs, so that any predators will be flooded with potential food. In a pond in Wales, for example, there was a dramatic decline in Common Toad (*Bufo bufo*) tadpoles as a result of diving beetle predation over just a few days. Nearly all diving beetle larvae visible in the water had a toad tadpole in their jaws. In such cases, even if only 2 percent of the tadpoles survive to metamorphosis this may be sufficient to ensure a stable population of toads. Conversely, if enough froglets are to make it to adulthood, some species that lay small clutches of eggs invest time and effort in caring for their offspring. About 10 percent of frog species show such behavior, which can take the form of building nests, attending to the eggs, carrying the eggs around, caring for the tadpoles (including transporting them to safer places), and even feeding the tadpoles. A relatively high proportion of frogs that breed in phytotelmata (water that accumulates within plants such as bromeliads and pitcher plants) show some degree of parental care (for more on this, see pages 44–45).

→ Many bufonid toads—such as the Cane Toad (*Rhinella marina*)—lay their eggs in long gelatinous strings that may be wound around rocks or vegetation.

EGG LAYING, NESTS, AND NURSERIES

EGGS, TADPOLES, AND PARENTING

Parental care

Parental care generally requires a greater investment in fewer offspring. Provisioning developing tadpoles with yolk to get them started in life can sometimes only go so far, and very few frog species have tadpoles that are sustained by a yolk supply alone, so at some stage most need an alternative food supply to complete their development. Some frog species even protect their eggs and developing young by swallowing them and keeping their offspring in a vocal sac or internal brood chamber.

→ In the Horned Land Frog (*Sphenophryne cornuta*) from New Guinea, fully metamorphosed froglets emerge from the eggs. These are carried on the back of the male for several days before they hop off and disperse into the forest.

← The female Palawan Wart Frog (*Limnonectes palavensis*) leaves her fertilized eggs with the male, who is dedicated to their care until they are transported to water as tadpoles.

FEEDING WITH INFERTILE EGGS

One way to provide further nutrition for the tadpoles is for the female frog to feed them with infertile eggs. Supplying tadpoles with infertile eggs was first observed in the tiny Strawberry Poison Frog (*Oophaga pumilio*, see page 128), which deposits its tadpoles in the water that accumulates in the leaf axils of bromeliad plants. The Mountain Chicken Frog (*Leptodactylus fallax*), from Montserrat and Dominica, is an order of magnitude larger than the Strawberry Poison Frog and builds a foam nest in a burrow on land. Here the tadpoles are entirely dependent on infertile eggs provided by the female. This reliance of the tadpoles on infertile eggs for food (known as obligate oophagy) is therefore an adaptation that allows reproduction to take place away from water bodies.

SWALLOWING EGGS

In addition to guarding or provisioning their tadpoles in a plant or nest, some frogs have evolved ingenious ways of caring for their offspring. Discovered by Charles Darwin in 1835, Darwin's Frog (*Rhinoderma darwinii*) is found in temperate forests of Chile and Argentina. After mating, the eggs are deposited in leaf litter. Although the female abandons the clutch, the male remains in close attendance. As the eggs develop, the tadpoles start wriggling inside the egg capsules. This movement stimulates the father to snap them up and swallow them, where they lodge in his vocal sac. Here they continue to develop for about two months. Once the yolk has been used up, the father releases a nutritious secretion into the vocal sac which continues to nourish the young. Completing metamorphosis entirely within the vocal sac, the young froglets move up into the mouth and are "burped out" by the male. The closely related Chile Darwin's Frog (*R. rufum*) also raises the eggs in the vocal sac, but releases the young tadpoles to complete their development in an external water body.

EGGS, TADPOLES, AND PARENTING

USING BROOD CHAMBERS

Discovered in 1972, the Southern Gastric Brooding Frog (*Rheobatrachus silus*) of Australia has one of the most unusual ways of caring for young in the entire animal kingdom. The female swallows her fertilized eggs into her stomach, which is converted to a brood chamber. The lining of the stomach goes through some structural changes to accommodate the eggs, which proceed to complete their full cycle of development into froglets. As the tadpoles develop and grow, the stomach becomes increasingly distended. When they are ready to leave the mother, they are regurgitated.

Intriguingly, the young release a chemical that neutralizes the naturally acidic stomach juices. This ability was of considerable biomedical interest at the time, as if the process could be simulated in humans, it could potentially be used to treat stomach ulcers. Sadly, however, along with its close relative, the Northern Gastric Brooding Frog (*R. vitellinus*), this remarkable amphibian became extinct in the 1980s, probably as a result of disease (see Chapter 9).

→ A newly metamorphosed Matang Narrow-mouthed Frog (*Microhyla nepenthicola*) emerges from the water of its pitcher plant home.

↓ As they metamorphose, African Bullfrog (*Pyxicephalus adspersus*) tadpoles start to develop a warty skin and a distinctive stripe down the back.

HATCHING OUT

How long it takes for tadpoles to hatch depends on a range of factors. Low temperature will slow development, but species that breed in temporary water bodies that start to dry out will hatch more quickly and have fast tadpole development. The African Bullfrog (*Pyxicephalus adspersus*, see page 130) goes a step further to ensure its tadpoles don't die in a drying pool, with the male excavating a channel to allow them to escape to a more permanent water body.

EGGS, TADPOLES, AND PARENTING

Tadpole development

The diversity of habitats in which frogs breed and lay eggs is reflected in the diversity of form and function in tadpoles. Some tadpoles feed on microscopic algae, while others swallow whole invertebrates. Some tadpoles are short and rotund and others resemble streamlined eels. In terms of size, the smallest tadpoles can be difficult to see with the naked eye, while the largest is the size of a small trout. In addition, tadpoles have a remarkable ability to alter their behavior and slow or speed up their growth in response to environmental pressures.

If we take a typical, pond-breeding frog or toad tadpole from the families Ranidae or Bufonidae as an example, the hatchling has no mouth or eyes and remains attached to the jelly capsule or an adjacent plant while it absorbs the remaining yolk from the egg. Close inspection will reveal three pairs of fluffy gills, but these rapidly become invisible as a gill flap—or operculum—grows over them to form an enclosed gill chamber. At the same time, the mouth and eyes develop, and the tadpole becomes a free-swimming animal capable of feeding independently.

The mouth is an intriguing structure with a horny beak surrounded by rows of tiny keratinized "teeth" and sensory papillae. Such is the variation in the arrangement of these structures in frog tadpoles that they have been used to identify and classify different species, which are otherwise very similar in appearance. Collectively, these structures comprise what is termed the "oral disc," and the mouthparts are used to scrape microscopic organisms from rocks and vegetation (sometimes referred to as "periphyton"), creating a suspension that is pumped into the mouth. The food particles are then either trapped on gill filters or in mucus and passed into the esophagus, while the water passes out through a short tube, or spiracle, on the operculum (depending on the species, the spiracle may be located on the right or left side or on the underside behind the mouth). Since vegetable matter is difficult to

TADPOLE DEVELOPMENT

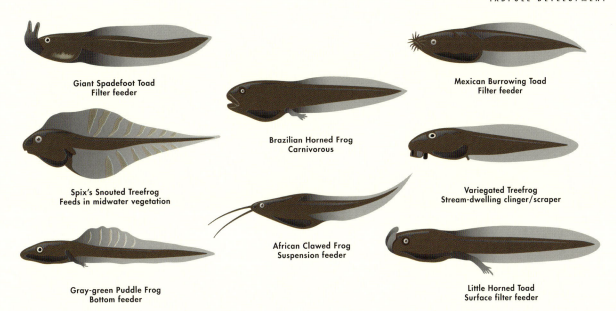

digest, the tadpole has a long, coiled gut that may be visible through the transparent skin on the underside. The mouthparts also allow tadpoles to scrape the carcasses of dead animals (including other tadpoles), creating a particulate soup that is passed over the internal filters.

Front and hind legs start growing at the same time, but, as the front legs are developing under the operculum, only the hind legs may be visible until quite late in the tadpole's development. The concealment of the front legs in the gill chamber and the growth of the hind legs beneath the tail reduces drag while the tadpole is swimming. The lungs usually start to function well before metamorphosis, and tadpoles can often be observed taking gulps of air from the surface.

Tadpoles
A range of tadpoles of different species showing the variation in mouthparts, body shape, and fin profile. These features are linked to the method of feeding and habitat selection.

↓ This Orange-legged Leaf Frog (*Phyllomedusa hypochondrialis*) shows several key stages in tadpole development and metamorphosis. As the legs develop and the body is remodeled into a terrestrial frog, the tail is resorbed, and the skin develops the adult pigmentation

EGGS, TADPOLES, AND PARENTING

Once the front legs burst through, the tadpole's body starts to go through some rapid remodeling, changing to something much more frog-like. The mouth widens and is restructured, so it can take live, invertebrate prey, and the gut shortens to accommodate the transition from a plant- to an animal-based diet. The eyes and head become more prominent, and, finally, the tail starts to be reabsorbed, leaving just a dark-colored tailbud between the hind legs, which are now fully grown and enable the newly transformed froglet to hop onto land with metamorphosis complete.

ADAPTATIONS FOR DIFFERENT HABITATS

There are many variations on this basic tadpole theme. The tadpoles of the African Clawed Frog (*Xenopus laevis*), for example, have translucent bodies and hang suspended in the water column. Instead of scraping mouthparts they have a pair of sensory tentacles and filter food particles from the water.

In tadpoles of megophryid frogs in Southeast Asia, many have funnel-like mouthparts for capturing tiny food particles trapped in the surface film of water. In one species, the Giant Horned Frog (*Atympanophrys gigantica*) of Vietnam, the mouthparts form four petal-like lobes that give the tadpole a plant-like appearance. At the other extreme, Budgett's Frog (*Lepidobatrachus laevis*, see page 182), from semiarid areas of South America, has a tadpole that is entirely carnivorous and even cannibalistic. Its jaws are structured much more like those of the adult frog and used to grab and consume whole prey, including other tadpoles.

KNOW YOUR SIBLINGS

When forming aggregations, tadpoles prefer to associate with siblings that emerged from the same clutch of eggs. As with schools of fishes, there are advantages to forming a group with similar-sized individuals, and these are more likely to be siblings that hatched at the same time. The reasons underlying this behavior are linked to the threat of being eaten and the chances of tadpoles transforming into frogs that can pass their genes on to the next generation. Evolutionary pressure means that those frogs with the "best" genes are more likely to survive. Siblings have more genes in common with each other than they do with non-siblings. Should a predator—like a fish or a dragonfly nymph—attack a shoal of tadpoles, then it is likely some will be eaten, but it is also probable that most will escape and survive thanks to safety in numbers. So, although those that die have made the ultimate sacrifice, some of their genes will be passed on through the survival of their siblings.

In stream-dwelling tadpoles the mouthparts take on a further role, forming a sucker that enables the animal to cling to rocks in flowing water. Tadpoles that live in very fast-flowing streams may even have an additional sucker behind the mouth, so nearly the whole underside of the animal can cling to a rock. Such is the strength of the sucker it has been reported that a tadpole of a Southeast Asian torrent frog (genus *Amolops*) could be held by its tail while remaining clasped to a rock many times its own weight.

TADPOLE BEHAVIOR

As is the case in all animals, the optimal behavioral strategy for tadpoles is one that maximizes growth and development while minimizing the risk of predation, competition, or contracting disease. This often involves aggregating in large numbers and dealing with competition from other tadpoles.

SAFETY IN NUMBERS

As growth and development is faster in warm water, tadpoles display clear daily cycles of behavior related to the thermal conditions of their water body. Many pond-breeding species, such as the European Common Frog (*Rana temporaria*, see page 32) and the American Toad (*Anaxyrus americanus*, see page 34), will move into sunny, shallow areas of water to enhance development, then disperse during hours of darkness. It can be difficult to tell whether this behavior is simply due to many tadpoles converging in an area of pond that enhances development or feeding, or to social attraction between tadpoles. However, many laboratory

↑ Tadpoles display complex behavioral interactions and can form structured "schools" and distinguish between their siblings and nonsiblings.

← Tadpoles of the Brazilian Horned Frog (*Ceratophrys aurita*) often eat other tadpoles, including their own kind.

between tadpoles. However, many laboratory experiments have shown that tadpoles are attracted to their own kind under controlled conditions, so some social behavior is certainly involved.

In such aggregations in shallow water, the collective wriggling of the tadpoles may stir up particulate matter and enable them to feed more efficiently as a group than as individuals. In deep water, tadpoles such as those of the European Common Toad (*Bufo bufo*, see page 104) may show a daily cycle of vertical movements in the water column, and this may be related to the movements of the planktonic organisms on which they feed.

Sometimes the aggregations are rather like those of fish in that the tadpoles are all swimming in the same direction. These polarized "schools" are particularly evident in species that swim in midwater, such as the tadpoles of the African Clawed Frog (*Xenopus laevis*). In addition to forming aggregations, in the presence of predators tadpoles can speed up their development, alter their swimming behavior to help them avoid detection, and even change shape and color to make them less likely to be eaten (see Chapter 6).

COMPETITIVE STREAKS

In small, temporary ponds the amount of food available may be in short supply, and there could also be several species of frog tadpoles competing for the same resources. Indeed, in the tropics 15 or more frog species may all breed in the same pond. When this occurs, the species may reduce competition by breeding at different times of year, or the tadpoles will use different parts of the same water body—some might stay within the vegetation while others use the shallows.

Nevertheless, there are situations when tadpoles will be using the same areas and feeding on the same food. In such cases, the general rule is that big tadpoles will have an advantage over small tadpoles. Large tadpoles may simply be able to exclude smaller tadpoles by virtue of their greater size. Moreover, as larger tadpoles have larger mouths, they will be scraping and filtering food in larger quantities than smaller tadpoles. Some may also release growth inhibitors into the water that slow the growth of smaller competitors. Whatever the mechanism involved, smaller tadpoles may take longer to reach metamorphosis and also metamorphose into a smaller frog, with associated implications for their longer-term survival.

The distribution of different species of tadpoles in relation to pond depth and vegetation
European Treefrogs (*Hyla arborea*) and Common Toads (*Bufo bufo*) are found at both the surface and the bottom, whereas Agile Frogs (*Rana dalmatina*) and Marsh Frogs (*Pelophylax ridibundus*) shift habitat use as they develop.

→ Large tadpoles can gain an advantage over small tadpoles by excluding them from food sources and by releasing growth inhibitors into the water.

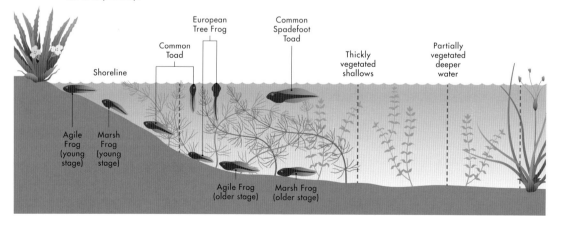

INHIBITING COMPETING TADPOLES

Large tadpoles can release cellular growth inhibitors into the water that slow the growth of smaller competitors. In Europe, Common Frogs (*Rana temporaria*, see page 32) sometimes breed in the same pools, but earlier in the year than Natterjack Toads (*Epidalea calamita*, see page 246). By the time the Natterjack tadpoles have hatched, the Common Frog tadpoles will already have a size advantage. However, as a lot of the Common Frog tadpoles may have died as a result of predation, there are likely to be rather fewer of these around by the time the Natterjack tadpoles start swimming and feeding. It is therefore in the interests of the Common Frog tadpoles to try and maintain the size advantage, so the natterjacks stay small and don't start to monopolize the pond's limited supply of algae and other microscopic organisms. They do this in a remarkable way. The large Common Frog tadpoles produce feces laden with a particular type of alga which small tadpoles find particularly attractive. This means the natterjacks are diverted away from eating the higher quality food toward the lower quality frog feces. In this way, the larger Common Frog tadpoles can inhibit the growth of the smaller Natterjack tadpoles and maintain their size advantage.

EGGS, TADPOLES, AND PARENTING

PIPA PIPA

Surinam Toad

Aquatic frog that broods young in pockets on its back

SCIENTIFIC NAME:	*Pipa pipa*
FAMILY:	Pipidae
LENGTH:	6–7 in (150–170 mm)
LIFE HISTORY:	Aquatic eggs (approximately 80 brooded in the skin) and adults; parental care
NOTABLE FEATURE:	Very flat body and a triangular head
IUCN RED LIST:	Least Concern

Also known as the Star-fingered Frog due to the curious, star-shaped organs on its toes used to detect food, this is one of the most highly specialized amphibians in the world in terms of general appearance (very flat, resembling a brown/olive-brown leaf), how it captures its food, and how it cares for its young.

An inhabitant of slow-moving rivers in South America, the Surinam Toad is also unusual in that it lacks a tongue. The star-shaped organs on the fingertips are sensitive to movements of food items in murky water, and the toad ambushes its prey by springing from a hiding place and opening its enormous jaws, so unsuspecting invertebrates or fish are sucked into the mouth, aided by the forelimbs.

Leaving home
When froglets of the Surinam Toad are ready to leave their individual pocket in the back of the female, they are around 2 cm (¾ in) long and immediately independent. Once all the young have left, the female sheds her skin and the now empty pockets on her back close up.

Froglet

To create extra suction, internal organs are displaced to the rear of the body to increase the current carrying the prey into the gullet.

The males don't croak like other frogs but attract females and warn off other males by clicking bones in the throat together. During mating, the male grabs the female around her waist in front of her back legs, which is known as "inguinal amplexus" (see page 94). The pair then perform a series of acrobatic somersaults in the water. During each somersault, the female releases a few eggs which are then fertilized by the male and pressed onto the back of his mate. The eggs sink into the female's skin and each egg becomes embedded in a fleshy pocket on her back. The tadpoles develop entirely in these pockets and eventually emerge as miniatures of the adults some weeks later. This leaves the female with a pock-marked back full of empty cavities, although these are lost when she sheds her skin. She is then ready to breed again.

There are currently eight species of toads within the genus *Pipa*, collectively known as Surinam toads. Although most of the species occur in Suriname, they are also widely distributed within the Amazonian region of South America.

→ After becoming embedded in her back, the eggs of the Surinam Toad are covered by a layer of skin. Here they spend the next 3–4 months developing directly into froglets.

OOPHAGA PUMILIO

Strawberry Poison Frog

Tiny frog that feeds its tadpoles with unfertilized eggs

SCIENTIFIC NAME:	*Oophaga pumilio*
FAMILY:	Dendrobatidae
LENGTH:	¾–1 in (17–24 mm)
LIFE HISTORY:	Aquatic eggs (up to 6) and tadpoles, terrestrial adults; parental care
NOTABLE FEATURE:	Several different color morphs from red to blue, spotted and unspotted
IUCN RED LIST:	Least Concern

The poison frogs of Central and South America (family Dendrobatidae) show some remarkable ways of caring for their eggs and tadpoles, but are perhaps better known for their bright colors and toxic skins. These toxins bioaccumulate in the skin because of the frogs' diet of ants and mites. Many species, including the Strawberry Poison Frog, are kept as pets, but in captivity they are not toxic as their diet consists of springtails, fruit flies, and small crickets, among other things.

In Strawberry Poison Frogs the males establish territories, which they defend against rival males while also attempting to attract females. Each male may mate with several different females and will guard the resulting clutches. As the eggs are laid in leaf litter, the male keeps them moist by watering them with the contents of his bladder.

The female is also territorial, and when her eggs hatch, she returns and carries one or more tadpoles on her back to a reservoir of water in the leaves of a bromeliad. Here she deposits a single tadpole, before returning to her clutch and transporting the rest of her offspring, one by one, to individual bromeliads. If more than one tadpole is deposited in a plant, only one will usually survive. Every few days, the female returns to each plant to feed the tadpoles with a few unfertilized eggs. Indeed, without this food the tadpoles are unlikely to survive. A recent discovery has revealed that the unfertilized eggs provide more than just food for the tadpoles. They are also loaded with chemicals that make the tadpoles toxic to eat, just like the adult frogs.

→ Strawberry Poison Frogs make excellent parents, but they are highly territorial and will actively remove the competition. If a male frog finds an unattended clutch of eggs that are not his own, he will eat them. Similarly, if the eggs are hatching, he will sit on them until a tadpole climbs on to his back—ordinarily this is the role of the female—and then deposit it in a water-filled bromeliad, thereby ensuring that the female will not know where her tadpoles are, and so won't be able to feed them.

EGGS, TADPOLES, AND PARENTING

PYXICEPHALUS ADSPERSUS

African Bullfrog

Large and feisty frog

SCIENTIFIC NAME:	*Pyxicephalus adspersus*
FAMILY:	Pyxicephalidae
LENGTH:	5½ to 9½ in (14 cm–24 cm)
LIFE HISTORY:	Aquatic eggs (up to 3,500 in clumps) and tadpoles, semi-aquatic adults; parental care
NOTABLE FEATURE:	Dominant males lek in the center of breeding territories
IUCN RED LIST:	Least Concern

Also known as the Giant Bullfrog, this is the largest frog in southern Africa, where it breeds in dry savannah areas with seasonal rainfall. This brownish-green frog has distinctive skin ridges on the back and a tubercle on each heel that is used for digging. The males are much larger than the females, highly territorial and aggressive, defending breeding pools against other bullfrogs and predators.

Breeding is initiated by heavy rains that fill pools and puddles. Smaller males congregate in communal areas around these pools, while the larger males chase away rivals and try and stop them breeding. When a female arrives she is usually grasped by one of the larger males, and mating ensues at the edge of the pool. The female then vacates the breeding pool, leaving the eggs and tadpoles in the care of the territorial male. The male is so aggressive that he will chase—and in some cases, even consume—smaller frogs and defend his offspring even when challenged by much larger predators such as herons. Male bullfrogs have even been known to attack humans (see pages 84–5)!

Dead bullfrogs have been found around the pools with multiple lacerations, indicating that they died in encounters with predators attempting to attack the offspring. However, their care of the young goes beyond seeing off all-comers. The breeding pools are small and shallow, and can reach temperatures of 104°F (40°C). If they become too hot or start drying out, the male bullfrog digs a shallow channel with his back legs to a deeper water body. These channels can be several feet long, and when they are complete the male ushers the developing tadpoles to safety along them. Along with their aggressive defense of eggs and tadpoles, this level of care from the male parent means that their offspring stand a much greater chance of survival.

→ African Bullfrogs are large anurans, with some reaching 3 lbs (1.4 kg) in weight. With their size, strong jaw muscles, and fanglike projections on the lower jaw, they are well-capable of taking small rodents as prey and can deliver a painful bite to any aggressor.

EGGS, TADPOLES, AND PARENTING

AMOLOPS LARUTENSIS

Larut Torrent Frog

The rock-clinging torrent frog

SCIENTIFIC NAME:	*Amolops larutensis*
FAMILY:	Ranidae
LENGTH:	1–2½ in (32–66 mm)
LIFE HISTORY:	Terrestrial eggs (on the undersurface of rocks over torrents) and adults; aquatic tadpoles
NOTABLE FEATURE:	Suctorial adaptations for life in fast-flowing water in frogs (toes) and tadpoles (ventral disc)
IUCN RED LIST:	Least Concern

The Larut Torrent Frog may be the most abundant frog living adjacent to the fast-flowing rivers and streams of eastern peninsula Malaysia. It is extremely well adapted to life in these torrents, with distinctive mottled and granular skin patterned to match its habitat, large eyes, and expanded suctorial disks on its toes.

Both adults and tadpoles are highly adapted to living alongside cascades and within fast-flowing water. Adults perch on rocks and vegetation, often vertically and facing head-down, from where they will readily dive into the torrential flows, emerging again shortly after. The toes of their feet have enlarged discs—digital pads—that allow these frogs to gain purchase on the slippery rocks in the splash-zone of torrential waters. Their vocalization is of a high frequency, which likely aids reception by other frogs above the sound of fast-flowing water, and they have very large eyes, suggesting that vision may play a greater role in (for example) male competition than in many other ranid frogs.

The most remarkable aspect of the Larut Torrent Frog concerns its tadpole, which has a distinctive morphology known as gastromyzophory. In short, the belly of the tadpole has become highly modified, and in this species has evolved to form a large disc-shaped sucker. This provides negative pressure, allowing the tadpole to "stick" to the rocky substrate and maintain its position while feeding on algae, or at rest in the fast-flowing water.

The Larut Torrent Frog was previously thought to occur across Thailand and throughout much of peninsular Malaysia, but genetic and morphological studies indicate that several species are present across the region, and that *Amolops larutensis* is in fact restricted to eastern areas of peninsular Malaysia. Intriguingly, the average body size of Malaysian torrent frogs appears to correlate with elevation, with larger frogs present in higher montane regions.

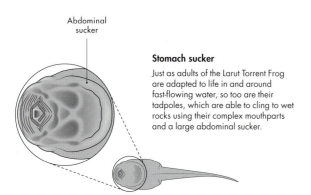

Stomach sucker
Just as adults of the Larut Torrent Frog are adapted to life in and around fast-flowing water, so too are their tadpoles, which are able to cling to wet rocks using their complex mouthparts and a large abdominal sucker.

→ With its specialized toe-pads and ultrasonic calls the Larut Torrent Frog is highly adapted for living alongside the loud torrential flows of fast-flowing rivers and streams.

EGGS, TADPOLES, AND PARENTING

LIMNONECTES PALAVANENSIS

Smooth Guardian Frog

Roles of males and females reversed

SCIENTIFIC NAME:	*Limnonectes palavanensis*
FAMILY:	Dicroglossidae
LENGTH:	1–1½ in (25–35 mm)
LIFE HISTORY:	Terrestrial eggs (clutch of 10–21) and adults; aquatic tadpoles
NOTABLE FEATURE:	Two fanglike projections on the lower jaw that are used for male combat
IUCN RED LIST:	Least Concern

The Smooth Guardian Frog has a dark band which runs horizontally between the eyes, dividing the different coloration of its snout and back. It also has a distinct V-shaped ridge between the shoulders, and some individuals have a white dorsal stripe. Occurring in the lowland forests of Borneo and the Philippine island of Palawan, this species is unusual in that, for any given population, the females outnumber the males.

Although many other species of frogs have males that look after the eggs and tadpoles, the Smooth Guardian Frog is unique in that the male forgoes other mating opportunities to look after a single clutch of his fertilized eggs. The time invested by male frogs in caring for their young means that the males are less available to female frogs looking for a mate. The resulting competition among females manifests in a lek-like congregation of calling females around a male frog. In contrast to the usual dynamic of lekking behavior, it is the male that responds to this attention, selecting a female with which to breed.

This sex role-reversal then extends to rearing the young. A successful female lays a clutch of eggs on the forest floor, which she then leaves with the male to care for. The male guards the eggs, protecting them from predators, during which time he may only feed on passing insects. When the eggs are ready to hatch—after approximately 10 days—the male crawls over the clutch and touches them with his chin and fingers. This helps to break the egg membranes, whereupon his tadpoles emerge and wriggle on to his back, a process that may take an hour or more. Up to 15 tadpoles have been observed riding piggy-back on the male, who then carries his offspring to a suitable water body, where he releases them to complete their development. Remarkably, when releasing tadpoles, male Smooth Guardian Frogs are known to divide their offspring between different pools—the anuran equivalent of not putting all your tadpoles "in one basket."

→ The male Smooth Guardian Frog not only guards his eggs. After they hatch, he carries them on his back to a suitable water body, where they are dropped off to complete their development.

GETTING AROUND

GETTING AROUND

Unique locomotion

The body plan of frogs is fundamentally different to that of most other four-legged vertebrates. Most mammals, many reptiles, and the majority of salamanders possess elongate bodies, four legs, and a tail. The legs on either side or under the body move in sequence and the tail may assist in balancing the body or even in locomotion itself. In contrast, frogs have a short, squat body and hind legs that are generally much longer than the front legs. The tadpole phase aside, they also lack a tail.

TYPES OF MOVEMENT

The fundamental frog body plan is adapted for dual purposes—hopping and leaping on land and swimming in water (see the diagram on page 140). In arguably the most familiar mode of locomotion, seen in frogs that divide their time between water and land, the long hind legs are used for launching the animal into the air from a standing (or rather squatting) start. In doing so, both legs push against the substrate together, and once off the ground the webbing and toes are folded in and the eyes closed to streamline passage through the air. Landing presents quite different challenges, and the short front legs and shoulder girdle are structured to act as shock absorbers when the frog hits the ground.

→ Frogs are remarkably agile, especially when in search of a meal, like this Pool Frog (*Pelophylax lessonae*) from northern Europe attempting to catch a dragonfly (*Libellula quadrimaculata*).

UNIQUE LOCOMOTION

GETTING AROUND

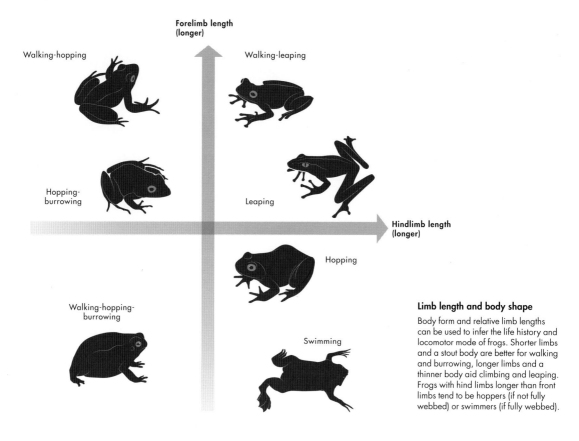

Limb length and body shape
Body form and relative limb lengths can be used to infer the life history and locomotor mode of frogs. Shorter limbs and a stout body are better for walking and burrowing, longer limbs and a thinner body aid climbing and leaping. Frogs with hind limbs longer than front limbs tend to be hoppers (if not fully webbed) or swimmers (if fully webbed).

In this way, many frogs can rapidly cover several feet in a series of repeated leaps. In water, the same hind legs are used for propulsion, with the front legs folded against the body to reduce water resistance. Interestingly, the ancient *Leiopelma* and *Ascaphus* frogs appear to have retained what might be the ancestral frog locomotory mode: when swimming they do so with alternate rather than synchronous thrusts of the hindlimbs, and when jumping—although now with synchronized hindlimb thrusts—tend to have an uncontrolled "belly flop" on landing, with the frog taking time to recover before the next leap. Similarly, Brazil's Pumpkin Toadlets (for example, *Brachycephalus pitanga*, see page 156) appear to have poorly developed balance systems, which seemingly results in uncoordinated jumps and crash landings.

The difference in length of the front and hind legs is rather less in the bufonid toads than in the ranid frogs. Consequently, many toads progress by a series of short hops rather than large leaps. Some frogs even walk by alternating the movement of the legs on either side of their body in the same way as lizards and salamanders. In arboreal frogs the hands have a gripping or adhesive function, while the flying frogs (Rhacophoridae) of Southeast Asia have extensive webbing on both fore- and hind feet that can act as parachutes or facilitate short glides (see Wallace's Flying Frog, *Rhacophorus nigropalmatus*, page 158).

Although anurans have a unique, highly conserved body plan, they display a wide variety of specializations across many different habitats which have resulted in observable differences in locomotion. Short forelimbs and hindlimbs indicate burrowing (and walking); long forelimbs and hindlimbs indicate arboreality (and leaping); hindlimbs longer than forelimbs indicates aquatic/semi-aquatic (more aquatic if heavily webbed) or hopper (more terrestrial if webbing is limited).

BREEDING MIGRATION AND FROGLET DISPERSAL

In both inward and outward migrations frogs may use a variety of cues to find their way to and from the breeding sites. For species such as many of the poison frogs (Dendrobatidae) that establish home ranges or defend territories on the forest floor, spatial awareness and movements may be guided by learned landmarks within those areas.

The other large-scale movement undertaken by frogs is when the young froglets disperse from the breeding site. Since the animals are tiny and difficult to track, little is known about their movements at this life stage, but it is likely they are particularly vulnerable to predators at this critical time. While dispersing, frogs may learn the locations of appropriate breeding ponds in relation to celestial and other environmental cues.

Navigating obstacles

The directional movements of frogs on land rarely occur in straight lines. As the structure of the landscape can vary, frogs may need to stick to areas with appropriate cover and humidity, such as hedgerows, forests, and long grass, and circumvent any inhospitable habitats that lie in their path. Nevertheless, if a road is built between the terrestrial habitat and breeding pond, migrating frogs and toads will still attempt to cross on damp nights and may consequently suffer high mortality from traffic. "Help a toad across the road" campaigns have subsequently become a popular way of engaging the public with amphibian conservation (citizen science) in several European countries.

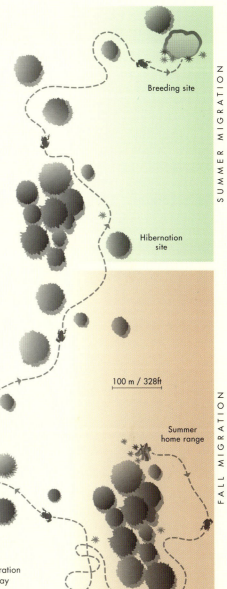

Seasonal movement
Seasonal movement of a male Common Toad (*Bufo bufo*) between the hibernation site, summer home range (for foraging) and breeding site.

← Barriers to traditional migration routes—such as new roads—may cause a problem for migrating toads. Toad tunnels built under roads have had mixed results in reducing road casualties.

GETTING AROUND

UNIQUE LOCOMOTION

← Wallace's Flying Frog (*Rhacophorus nigropalmatus*) uses extensive foot-webbing to aid maneuverability as it parachutes and glides from tree to tree.

REASONS FOR MOVING

Compared to mammals and birds, frogs do not tend to move particularly far in their daily lives. In fact, some frogs—such as those that live in plants like the Itambé Bromeliad Frog (*Crossodactylodes itambe*)—spend their entire lives in an area no more than about ten square feet or so. Likewise, even during the active season frogs may spend many days in a terrestrial hiding place rather than venturing out.

Unlike many other animals that have strong biological rhythms linked to the daily cycle of light and darkness, the movements of frogs are more strongly tied to environmental conditions such as temperature, rainfall, and humidity. However, some species undertake yearly nighttime migrations to breeding ponds that can be quite spectacular in terms of the numbers of animals on the move and the distances they cover. European Common Toads (*Bufo bufo*, see page 104) may migrate distances of 2 miles (3 km) or more to reach their breeding sites. Green Frogs (*Lithobates clamitans*, see page 160) stay in the vicinity of their water bodies all year round but make forays on to land to find food and hibernation sites. Rather less is known about how arboreal species move within trees and bushes, but in flooded forests at least it is likely that many frogs move up and down the vegetation according to prevailing water levels.

↑ A typical shallow breeding pool of the Natterjack Toad (*Epidalea calamita*) in a sand dune system. The size and location of such breeding pools may shift as the dunes respond to the impact of winds.

↗ The European Green Toad (*Bufotes viridis*) is able to take advantage of ponds created in highly disturbed habitats.

COLONIZING NEW BREEDING SITES

A common misconception is that all frogs return to breed in the place where they were spawned. Although many (indeed, probably the majority) return to their natal pond, many certainly don't, as this would not explain how quickly newly created or restored habitats are colonized. In North America's Wood Frog (*Lithobates sylvaticus*, see page 72), for example, about 18 percent of metamorphs disperse to different sites from those in which they were spawned. As a general rule for most species, the closer a new pond is to another breeding area, the quicker it is likely to be colonized. Although some adult frogs move between different breeding sites, they will in all probability return to a pond they have used in previous years. Indeed, there are frequent accounts

of frogs migrating back to breeding sites that have been destroyed or queuing up at barriers erected in their path.

That said, some frogs are more pioneering than others. The Natterjack Toad (*Epidalea calamita*, see page 246), for example, is able to use the shallow slacks of water that form between sand dunes. As the dunes shift, so may the water bodies, but the toads will move with them. Likewise, the Green Toad (*Bufotes viridis*) often moves into areas that have been opened up due to anthropogenic disturbance and may therefore be found in towns and cities in Europe. Other species are able to track more severe disturbances such as floods and wildfires to find the best breeding sites (see the Western Toad, *Anaxyrus boreas*, page 162).

Molecular genetics studies have shown that the flow of genes between populations is rather larger than direct measures of the journeys made by frogs might suggest. This may simply be down to the methodological constraints of trying to track the movements of thousands of tiny, cryptic animals. Consequently, dispersal of frogs between populations may be higher than we can directly measure.

GETTING AROUND

On the move

Frogs are masters at conserving energy and many species spend long periods of time not moving around at all. During these periods, they remain inactive in a moist hiding place, waiting for favorable conditions to return for breeding and feeding. When they do emerge to move toward their breeding areas or establish home ranges and territories, they use a variety of cues to orientate themselves within their environment.

A simple mechanism used by frogs to find their way around is sometimes known as "beaconing." This involves heading toward a distant cue, such as a chorus of calling frogs or the odor of a pond. Alternatively, a frog may learn local landmarks within its home range and use these to guide its movements. Unless it can draw on other information, should a frog find itself outside the area with which it is familiar, it may effectively be lost and unable to find its way.

COMPASS ORIENTATION

Compass orientation has been demonstrated in a wide range of amphibians and involves using the Sun—or possibly other celestial cues like the Moon and stars—to move around. This means learning where a pond or the land lies in relation to the position of the Sun, for example. However, as the day progresses, and the Sun moves across the sky, the orienting frog must adjust its angle of movement accordingly. So it uses an internal biological clock to tell it where the Sun should be at a certain time of day.

← The Southern Cricket Frog (*Acris gryllus*) has strong homing skills and is able to use celestial cues to find its way around.

ON THE MOVE

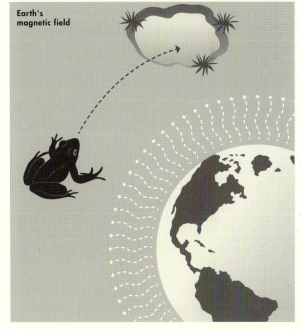

Navigation

Frogs use a variety of cues to traverse their habitat. These include "beaconing," whereby they respond to a directional signal, such as the calls of other frogs; "landmarks," which are key features of the habitats where they live; "compass orientation," whereby they learn and maintain a certain angle in relation to the Sun, stars, or Moon; and "environmental cues," such as the odor of a pond or the Earth's magnetic field.

North American cricket frogs (*Acris crepitans* and *A. gryllus*), for instance, learn where the nearest shoreline is in relation to celestial cues and swim in the right direction when placed in enclosures with only the sky visible.

Such compass orientation has also been observed in froglets of several other species when emerging from a pond to find their way toward an appropriate terrestrial habitat. However, if those frogs that rely on compass orientation are displaced a certain distance from the pond, they will move in a direction parallel to the original direction of travel (that is, they are unable to compensate for the displacement). In contrast, true navigation involves working out the exact current location even if displaced and adjusting the direction of movement accordingly.

OTHER NAVIGATION METHODS

The reality is that frogs probably use a range of environmental cues to find their way around (see page 147). Some species—including the European Common Frog (*Rana temporaria*, see page 32)—have been shown to respond to the odors associated with their breeding ponds, and finding the pond can be helped by the wind blowing the odors in the right direction. But odor is unlikely to be the only cue used. As in migrating birds, there is also evidence that frogs may be able to use the Earth's magnetic field to orientate themselves. The direction of breeding migrations of the Marsh Frog (*Pelophylax ridibundus*) and the European Common Toad (*Bufo bufo*, see page 104), for example, can be shifted by placing the animals in artificial magnetic fields. This may be useful when frogs are unable to use the Sun, Moon, or stars as a compass because the celestial cues are masked by cloud cover.

← ↖ Both Marsh Frogs (*Pelophylax ridibundus*; top) and Common Toads (*Bufo bufo*; bottom) may use the Earth's magnetic field during breeding migrations.

ON THE MOVE

INVASIVE CANE TOADS

Originally native to South America, Cane Toads (*Rhinella marina*) were released into northeastern Australia in 1935, supposedly to control insect pests. The toads have rapidly expanded in both westerly and southerly directions, and their toxic skins have had a devastating impact on native species that attempt to eat them. Radiotracking research has shown that Cane Toads expanding their range into tropical Australia move farther each day than they do in areas where they are native or already well-established. In fact, pioneering toads at the invasion front can move up to 656 ft (200 m) per night. They do this by traveling farther each night and changing daytime hiding places less frequently. This results in a strong selection pressure at the leading edge of the range, and the evolution of toads that can move faster and farther. Under warm, wet conditions then, Cane Toads can use this mechanism to expand the front of their invasion at a rate of over 30 miles (50 km) per year.

Rapid expansion

Following its introduction to Australia in 1935, the Cane Toad has expanded its range across more than 500,000 square miles (1.3 million square km) of northern and eastern Australia. Individual toads can disperse more than 1 mile (1.8 km) each night, and use roads as dispersal corridors.

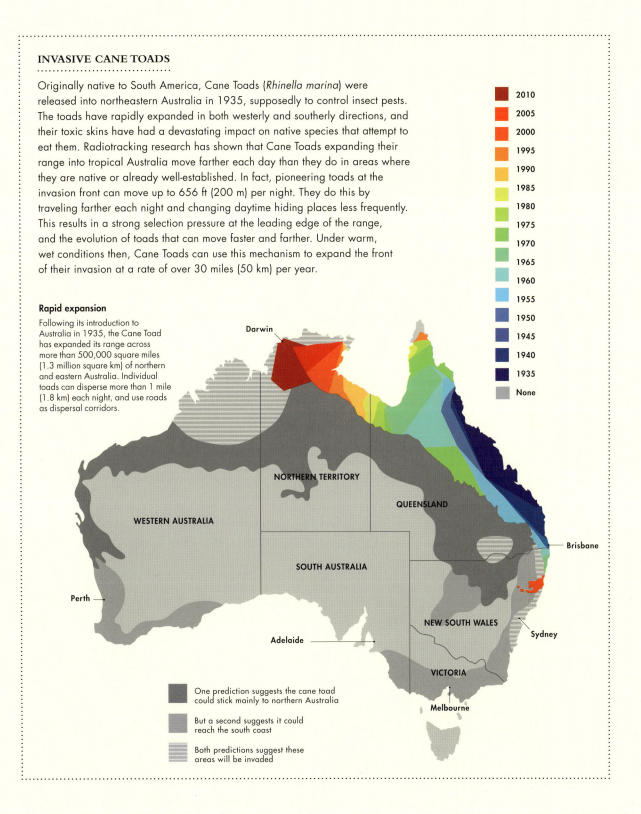

FINDING THE WAY HOME

What is clear is that frogs can learn to find their way around their immediate environment and even their way back if they are displaced. In particular, the poison frogs of South America (Dendrobatidae) have a life history that requires good spatial awareness because they establish territories on the forest floor. Some species transport tadpoles from nests to water bodies, and others return regularly to those water bodies to feed their young. Radio-tracking studies of the Three-striped Poison Frog (*Ameerega trivittata*) and the Brilliant-thighed Poison Frog (*Allobates femoralis*) have shown that they can identify the most direct route back when displaced to areas outside those used for routine movements. Three-striped Poison Frogs are even capable of navigating to their home territory from distances of nearly half a mile (792 m) away.

Other work on the Green and Black Poison Frog (*Dendrobates auratus*) has shown that they can use visual cues to work out routes from multiple different positions, suggesting that, as in other vertebrates, they are able to compile a "cognitive map" of their surroundings. Although frogs can certainly learn landmarks to find their way around familiar areas, we do not yet fully understand how some are able to navigate home from more distant, unfamiliar areas. In such cases, it is possible that the cues they use have yet to be discovered.

↓ Male Dyeing Poison Frogs (*Dendrobates tinctorius*) fitted with radio transmitters have been tracked transporting single tadpoles to favored water bodies over hundreds of meters, and have good spatial and location memory within their home range.

→ Another dendrobatid, the Brilliant-thighed Poison Frog (*Allobates femoralis*) transports multiple tadpoles at a time, and radio-tracking studies on this species indicate that extensive spatial learning based on landmarks helps these frogs to navigate between favored pools for tadpole deposition. Olfactory cues may also play a role, in particular when locating new sites.

GETTING AROUND

Alien invaders

Humans have often transported frogs with them when moving around the world. Sometimes they are inadvertent stowaways among transported goods, or they are deliberately released to control pests, or to be used as a food source. Either way, the impact of non-native frogs on native species can be far-reaching.

→ The Cane Toad (*Rhinella marina*) has been introduced to several countries around the word, primarily to control agricultural pests (in which it has largely failed).

↙ The Lesser Antillean Whistling Frog (*Eleutherodactylus johnstonei*) is widespread in the Caribbean and a good natural disperser, it is also easily concealed in goods transported between islands.

When humans travel around the world they invariably take a range of other species with them, and this includes frogs. Although vulnerable to temperature extremes and desiccation, frogs can exist without food for long periods and are therefore able to stowaway among vegetables, ornamental plants, soil, or other materials being transported long distances. In some cases, such as that of the Lesser Antillean Frog (*Eleutherodactylus johnstonei*, see page 164), it is difficult to distinguish between dispersal by natural means and dispersal by human agency.

Although native to the eastern United States, the American Bullfrog (*Lithobates catesbeianus*, see page 96) was extensively released into western states as a source of food for miners and settlers. The same species has also established breeding populations in several European countries and parts of Asia after being imported for farming or bred for the pet trade. Whether initial releases were deliberate or accidental, the American Bullfrog is a voracious predator, keen competitor, and also a vector for the chytrid fungus, which can kill other amphibians. As a result, it has been able to expand its introduced range through natural dispersal and proven difficult to control. A similar issue applies to the Cane Toad (*Rhinella marina*), which is expanding its range in Australia (see page 149). If the climatic conditions in a new area are conducive to

breeding and dispersal, when a non-native species is introduced outside its natural range it may be able to disperse much faster than within its natural range.

DETRIMENTAL EFFECTS

Overall, the impacts of humans moving frogs around the world—either deliberately or accidentally—are rather mixed. Many frogs die out soon after release, some persist locally for several generations without any significant impact on native species, while others can have severe impacts on other species and habitats that are difficult to mitigate. What is of more concern is the potential for the movement of amphibians around the world to transmit infectious diseases, such as those caused by the chytrid fungus that is wiping out frog populations and species in different parts of the globe (see Chapter 9). Where pathogens have co-evolved with their natural hosts frogs can develop a natural immunity. However, when those frogs are moved to another part of the world and come into contact with other amphibian species, the pathogens can jump hosts and cause disease in animals that have no natural resistance.

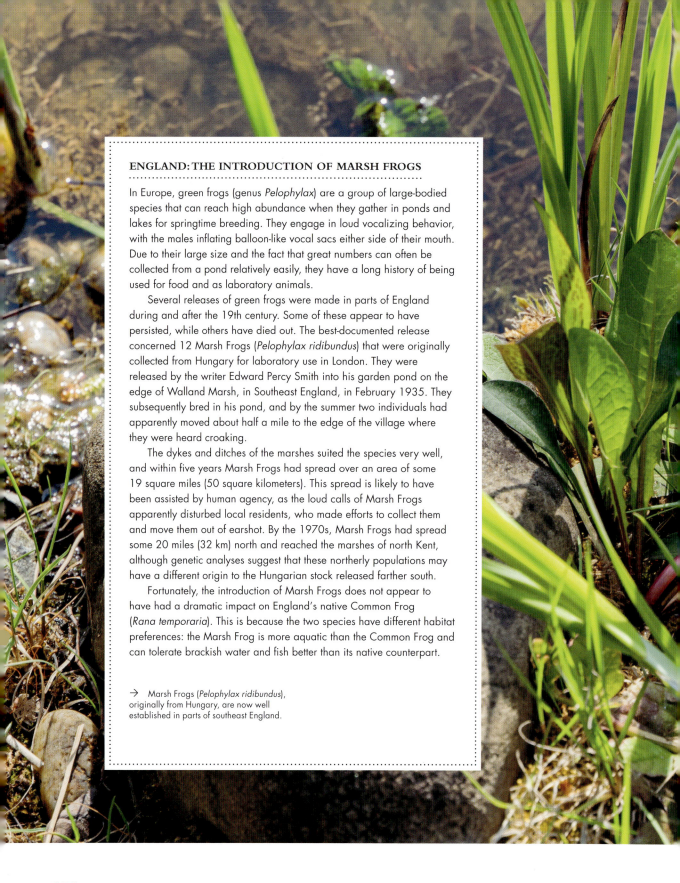

ENGLAND: THE INTRODUCTION OF MARSH FROGS

In Europe, green frogs (genus *Pelophylax*) are a group of large-bodied species that can reach high abundance when they gather in ponds and lakes for springtime breeding. They engage in loud vocalizing behavior, with the males inflating balloon-like vocal sacs either side of their mouth. Due to their large size and the fact that great numbers can often be collected from a pond relatively easily, they have a long history of being used for food and as laboratory animals.

Several releases of green frogs were made in parts of England during and after the 19th century. Some of these appear to have persisted, while others have died out. The best-documented release concerned 12 Marsh Frogs (*Pelophylax ridibundus*) that were originally collected from Hungary for laboratory use in London. They were released by the writer Edward Percy Smith into his garden pond on the edge of Walland Marsh, in Southeast England, in February 1935. They subsequently bred in his pond, and by the summer two individuals had apparently moved about half a mile to the edge of the village where they were heard croaking.

The dykes and ditches of the marshes suited the species very well, and within five years Marsh Frogs had spread over an area of some 19 square miles (50 square kilometers). This spread is likely to have been assisted by human agency, as the loud calls of Marsh Frogs apparently disturbed local residents, who made efforts to collect them and move them out of earshot. By the 1970s, Marsh Frogs had spread some 20 miles (32 km) north and reached the marshes of north Kent, although genetic analyses suggest that these northerly populations may have a different origin to the Hungarian stock released farther south.

Fortunately, the introduction of Marsh Frogs does not appear to have had a dramatic impact on England's native Common Frog (*Rana temporaria*). This is because the two species have different habitat preferences: the Marsh Frog is more aquatic than the Common Frog and can tolerate brackish water and fish better than its native counterpart.

→ Marsh Frogs (*Pelophylax ridibundus*), originally from Hungary, are now well established in parts of southeast England.

GETTING AROUND

BRACHYCEPHALUS PITANGA

Red Pumpkin Toadlet

Tiny toxic frog

SCIENTIFIC NAME:	*Brachycephalus pitanga*
FAMILY:	Brachycephalidae
LENGTH	½–¾ in (11–14 mm)
LIFE HISTORY	Terrestrial eggs (direct development, clutch size unknown but probably around five eggs) and adults; parental care
NOTABLE FEATURE	A head and upper skeleton that fluoresces under ultraviolet light
IUCN RED LIST	Least Concern

Brachycephalus pitanga belongs to a group of tiny frogs known as pumpkin toadlets. Found in a small region of Brazil's Atlantic Forest, the Red Pumpkin Toadlet has a bright orange background coloration dotted with contrasting patches of red, especially on the back, which warns predators that they are not good to eat. Their diminutive stature seems to have brought several challenges to their behavior and morphology.

In the inner ear of vertebrates lies a structure known as the semicircular canals. These are a set of tiny bony tubes, within which the flow of fluid helps the animal to maintain its balance and posture. Pumpkin Toadlets have the smallest known semicircular canals of any vertebrate, suggesting an imperfect sense of balance. This is not a problem when a Pumpkin Toadlet moves around using all four limbs, but when it jumps it has very little control over its flight through the air or how it lands. After propelling itself forward in a jump, the toadlet body may cartwheel and spin unpredictably, resulting in an uncoordinated crash landing. Consequently, the toadlets sometimes land on their head or back and may take a few seconds to recover before righting themselves.

Although Pumpkin Toadlets vocalize, they are unable to hear each other's calls, likely due to the incomplete development of the inner ear. It is possible that the visual signals associated with the calls are more important than the sounds themselves—for example, male frogs are known to wave their forelimbs in territorial displays. Ordinarily, frog vocalizations carry a risk to the caller as they can alert predators to their presence. However, in the case of Pumpkin Toadlets, their bright coloration warns predators of their toxicity, and it is perhaps because of this there has been no evolutionary pressure for these frogs to cease producing audible sound.

Whereas some frogs have osteoderms—bony plates in the skin—the skin on the head and back of the Red Pumpkin Toadlet is co-ossified, meaning that the skin and skeleton are essentially fused in these regions. A further morphological feature of the Red Pumpkin Toadlet (and other *Brachycephalus* frogs) is the loss of bones in the hands and feet. Extensive ossification, and loss and/or reduction of skeletal elements (for example, the inner ear and limbs of Pumpkin Toadlets) are generally considered an evolutionary consequence of miniaturization.

→ The distinctive bumpy appearance of the skin on the head and back of Red Pumpkin Toadlets is due to the skin being formed with a matrix of bone.

GETTING AROUND

RHACOPHORUS NIGROPALMATUS

Wallace's Flying Frog

Rainforest glider

SCIENTIFIC NAME:	*Rhacophorus nigropalmatus*
FAMILY:	Rhacophoridae
LENGTH:	Up to 4 in (100 mm)
LIFE HISTORY:	Terrestrial eggs (foam nest, clutch size possibly up to 100 eggs) and adults; aquatic tadpoles
NOTABLE FEATURE:	Rarely seen, except for when it descends from the canopy to breed
IUCN RED LIST:	Least Concern

Named for Alfred Russel Wallace, who along with Charles Darwin helped frame the theory of evolution by natural selection, Wallace's Flying Frog is found in the forests of Southeast Asia. Extensive webbing and skin flaps assist the frog in gliding from tree to tree in a controlled descent. Wallace's Flying Frog is vivid green with orange and/or yellow fingers and toes connected by distinctive black webbing.

Although gliding has evolved in treefrogs from different families, it is particularly well developed in several species of the genus *Rhacophorus*. Some frogs simply spread out their legs and use their often limited webbing to parachute to the ground as a way of reducing impact. However, Wallace's Flying Frog enhances maneuverability as it descends by banking its body and adjusting the positions of its legs.

Using fully webbed hands and feet, together with flaps of skin running along the margins of their limbs, these frogs can control their descent through complex forest environments, and adjust the horizontal distance covered during the glide. Changing their body positions while in the air further enhances this ability.

Expanded toe pads on their enlarged hands and feet also seem to have a stronger adhesive ability than those of other frogs, which may be important when landing on either rough, vertical surfaces like tree trunks, or smooth flat surfaces, such as on leaves.

When young, Wallace's Flying Frog has an orange-red coloration, and this changes to a more typical emerald green hue as the frogs reach adult size. Carotenoids (pigment molecules) are thought to play an important role in this process of color change during growth and development. The frog's ability to rapidly change its adult coloration to lighter or darker green is also attributed to carotenoids, something that allows it to camouflage itself in the lighter canopy or dark understory of its rainforest home.

How "flying" frogs glide
When moving from tree to tree through the rainforest canopy, Wallace's Flying Frog not only manages different eddies of airflow (shown by the light-gray arrows) by adjusting its body position (the darker arrows), it can also combine pitch, yaw, and roll (as do birds, and airline pilots!) to direct and control its glide toward landing.

→ Wallace's Flying Frog does not fly in the strictest sense, but it is arguably the most accomplished glider of all frogs that do so.

GETTING AROUND

LITHOBATES CLAMITANS

Green Frog

A vocal and territorial frog

SCIENTIFIC NAME:	*Lithobates clamitans*
FAMILY:	Ranidae
LENGTH:	2¼–4¼ in (57–108 mm)
LIFE HISTORY:	Aquatic eggs (up to 7,000 in clumps) and tadpoles, semiaquatic adults
NOTABLE FEATURE:	Size and color vary; two subspecies are recognized, *L. clamitans clamitans* and *L. c. melanota*.
IUCN RED LIST:	Least Concern

The North American Green Frog is a large, widespread, and often abundant species with a broad distribution in eastern North America, ranging from southern Canada to northern Florida. Its coloration varies from browns and greens (often both), or sometimes bluish. Green Frogs generally have a dark fold running down the back from behind each eye and barred patterning on the limbs.

Green Frogs use a variety of wetlands that hold water all year round and lay their eggs in shallow water among reeds and rushes. Unlike some other frogs, the males remain in the vicinity of the water when breeding is over. Indeed, Green Frogs do not display the large-scale migrations to and from breeding areas that is often seen in other related species.

Radiotracking work on Green Frogs in New York State has shown that during the fall some individuals make forays away from breeding sites to feed terrestrially. As the abundance of food increased at a greater distance from the ponds, those frogs that foraged farther away increased their body weights. Nevertheless, the foraging frogs regularly returned to the ponds where they would be safer from predators, and where the warmer water (which retains heat longer than the surrounding air) can aid digestion.

Toward the end of fall, Green Frogs make separate journeys to seek out hibernation sites in the form of streams and ground-water seepages. Overwintering in these areas may be safer than at the bottom of a pond, which can become covered in ice in winter with any hibernating frogs running the risk of suffocation due to deoxygenation. Depending on the environmental conditions, in other areas Green Frogs may hibernate on land.

→ North American Green Frogs tend to have an extended breeding period that varies from March through to September depending on latitude—starting earlier and lasting longer in the south. While breeding, male North American Green Frogs are territorial and defend their favored patch with a call that is likened to a plucked banjo string!

GETTING AROUND

ANAXYRUS BOREAS

Western Toad

Breeds in montane snowmelt ponds as well as desert streams

SCIENTIFIC NAME:	*Anaxyrus boreas*
FAMILY:	Bufonidae
LENGTH:	Up to 5 in (130 mm)
LIFE HISTORY:	Aquatic eggs (up to 12,000 in strings) and tadpoles, terrestrial adults
NOTABLE FEATURE:	Light-colored stripe down the middle of the back
IUCN RED LIST:	Least Concern

The Western (or Boreal) Toad has a wide distribution in western North America, ranging from southeast Alaska to New Mexico. Western Toads display a typical toad breeding cycle, with adults hibernating on land and migrating to breeding ponds. Unusually, however, the males lack vocal sacs, although they can still produce weak calls. The Western Toad's coloration varies from gray to greenish with dark blotches, often with reddish patterning.

Recent research in Wyoming has demonstrated that Western Toads can adjust their movements in response to unpredictable environmental changes. In 2017 extreme flooding breached several beaver dams, which meant that the deep, sparsely vegetated pools where the toads prefer to breed were converted to less suitable shallow, vegetated pools. However, the toads were able to track the change in habitat and move from the shallow pools—where metamorphosis was likely to be less successful—to other deeper pools.

The same researchers observed a similar response to another unpredictable event—wildfire. In 2018 a wildfire burned nearly 62,000 acres (25,000 hectares) of land around creeks where Western Toads bred. Two years after the fire, toads appeared to disperse from burned areas and resettle in unburned areas. Interestingly, the growth of toads appeared to be better in the burned areas, possibly because invertebrate food increased after the wildfire. Despite the severity of the fire, there was no apparent impact on the survival of adult toads or the recruitment of toads to the adult population. This example illustrates how toad life histories are adapted to deal with unpredictable events and shifting landscapes. With up to 20 percent of the toads regarded as nomadic, dispersal to new areas in the event of a disaster may help to sustain the population within the wider landscape.

→ A widespread and generally adaptable species, the home range of female Western Toads can be nearly 42 acres (17 hectares), and they may travel as much as 2.5 miles (4 km) to reach a yearly breeding site. Male toads tend to have a smaller home range, and travel shorter distances.

GETTING AROUND

ELEUTHERODACTYLUS JOHNSTONEI
Lesser Antillean Frog
This frog can whistle 120 times a minute

SCIENTIFIC NAME:	*Eleutherodactylus johnstonei*
FAMILY:	Eleutherodactylidae
LENGTH	Up to 1¼ in (35 mm)
LIFE HISTORY	Terrestrial eggs (10–30 direct developing) and adults; parental care
NOTABLE FEATURE	Its loud call makes this a frog more often heard than seen
IUCN RED LIST	Least Concern

The genus *Eleutherodactylus* comprises over 200 species of small, cryptic frogs distributed across southern North America and the Neotropics, including the Caribbean. *E. johnstonei* is the species that serenades many tourists with its choruses in the Caribbean at dusk and dawn. Also known as Johnstone's Whistling Frog, it has an enigmatic distribution across the islands of the Lesser Antilles, but also reaches northern South America, Jamaica, and Bermuda. Its coloration ranges from light brown to gray, usually with one or two darker, V-shaped markings.

There has always been a high degree of uncertainty about where this species is native and where it has been introduced. There are certainly well-documented accounts of it being deliberately introduced by people into a number of areas, including Jamaica and Bermuda. Given its small size and tendency to hide among leaves, it is also highly likely that the Lesser Antillean Frog has moved around as a stowaway in materials being transported between islands. In the 19th century, some frogs even became established in London's Kew Gardens from plants transported to one of its greenhouses, and although the species was originally described from specimens from Grenada, it is unlikely it was ever native there. Moreover, the Lesser Antillean Frog is also a resilient competitor, well able to colonize disturbed habitats of its own accord and tolerant of temperature fluctuations. In addition, as a direct developer, it does not require standing water to breed.

Adding to the confusion over its origin, it has often been misidentified as another whistling frog—*Eleutherodactylus martinicensis*. Unscrambling the origins of the distribution pattern and how far these frogs might actively disperse has had to await the arrival of molecular methods that can be applied to specimens collected from across the range.

Recent genetic analyses have shown that the Lesser Antillean Frog is native to the island of Montserrat. The common ancestor of the species on Montserrat dates from 0.24–0.81 mya, well before any human colonization of the region. The species exists in distinct eastern and western geographic clades. Although the source of the eastern clade is uncertain, it is frogs from this clade that appear to have been widely distributed elsewhere.

→ Its small size means the Lesser Antillean Frog can stowaway in plants transported by the nursery trade. On islands where it has been introduced, it poses a risk to native frog species— including other eleutherodactylids— by becoming a competitor for resources such as food and habitat.

FROGS, FOOD, & FEEDING

FROGS, FOOD, AND FEEDING

A variety of diets

The majority of frogs are generalist predators and can be described as insectivorous or carnivorous, taking prey that largely consists of invertebrates and small vertebrates. Many frogs even eat other amphibians, including other frogs, and some will readily cannibalize each other. There are also frogs that feed on carrion, and, although uncommon, at least two species have adapted to seek out and include plant and plant-derived material in their diet to such an extent that this contributes a significant proportion of their regular subsistence.

Although numerous frogs will consume any moving prey they encounter of an appropriate size (that is, anything mouth-sized and swallowable), many are highly specialized. Small species of frog (and indeed newly metamorphosed frogs of larger taxa) feed on specific types of prey based on their size, such as the mites and ants that form the diet of the Gardiner's Seychelles Frog (*Sechellophryne gardineri*), which is just under half an inch (11 mm) long. In the case of *Paedophryne amauensis* —one of the world's smallest frogs at a third of an inch (7.7 mm)—the prey is springtails, themselves often less than 2 mm in length. However, true specialisms are also abundant across the anuran tree of life,

← Cane Toads (*Rhinella marina*) have cannibalistic tendencies from an early age, and larger individuals (like this juvenile) will often prey on smaller toads.

→ Environmental conditions and prey availability can influence dietary breadth or specificity, and having a wide prey-base provides the opportunity for Cane Toads to take advantage of anything that comes along, whether that's a relative, or another vertebrate such as a caecilian (in this case, *Caecilia tentaculata*).

FROGS, FOOD, AND FEEDING

← Many frogs will also feed on other frogs—such as this Giant River Frog (*Limnonectes leporinus*), which is in the process of swallowing a Mahogany Frog (*Abavorana luctuosa*).

Specialists in slugs and snails

There are only two species in the genus *Paracassina* and both are mollusk specialists. The Ethiopian Striped Frog (*Paracassina obscura*) has a short, robust skull and favors snails which it swallows whole, shell and all. Although its sister species, the Kouni Valley Striped Frog (*Paracassina kounhiensis*) will also take smaller snails, it has long recurved teeth on the jaw making it adept at securing its chosen prey of slugs.

Ethiopian Striped Frog

Kouni Valley Striped frog

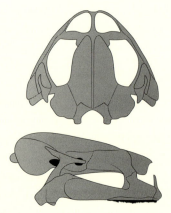

and while the overall size of a frog does not always correspond with the size of its prey, the size of its head and feeding apparatus regularly does.

While some frogs are highly specialized for a given diet—for example, slugs, in the case of the Kouni Valley Striped Frog (*Paracassina kounhiensis*), others will feed on seemingly uncharacteristic prey items when the opportunity presents itself. This is the case for the West African Brown Frog (*Aubria subsigillata*), which reportedly catches fish if they leap from the water. Among anurans, the larger species of bufonid toads appear to have the greatest propensity for expanding their dietary intake. The Cane Toad (*Rhinella marina*) will readily consume dry and wet pet food left outside for dogs and cats, while the Southern Toad (*Anaxyrus terrestris*) has been seen feeding on road-killed River-swamp Frogs (*Lithobates heckscheri*). The Cururu Toad (*Rhinella diptycha*) probably wins out though—its feeding behavior can at times be described as cannibalistic necrophagy, on account of its feeding on road-killed relatives.

A VARIETY OF DIETS

BODY SIZE, SKULL SIZE, AND DIET

Many frogs are adapted to feed on particularly small prey despite being orders of magnitude larger than their target food. Adaptations such as a short, pointed snout and relatively tall skull are ideal for consuming ants and termites (myrmecophagy), and several species, in particular burrowing ones, have convergently evolved this distinctive skull morphology. Examples include the Mexican Burrowing Toad (*Rhinophrynus dorsalis*, see page 180) and Africa's Guinea Snout-burrower (*Hemisus guineensis*). As such, there is a general correlation between skull and jaw size and shape, so frogs which feed on small prey tend to possess narrower heads and shorter jaws and require neither a particularly wide gape nor strong bite force— all adaptations which confer a short period between capturing and ingesting prey.

Conversely, frogs that feed on larger prey— which necessarily require greater effort and take longer to consume than smaller prey—will have wider heads, longer jaws, a wider gape, and a greater bite force. However, being a large frog confers no disposition toward only eating larger prey. Although the larger frog species tend to eat the largest prey (including other vertebrates), they are often generalist and opportunistic feeders, with a greater dietary breadth and able to consume a greater size range of prey items than smaller, more specialist frogs. Larger species also tend toward greater ossification of the skull, making it more robust.

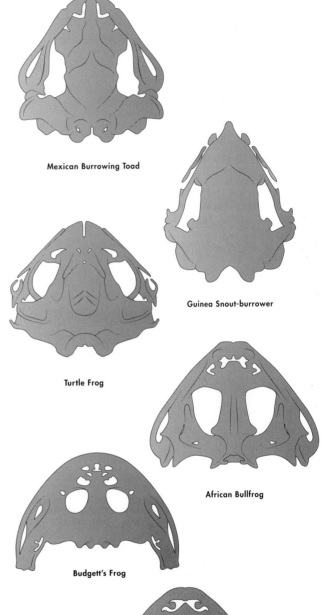

Mexican Burrowing Toad

Guinea Snout-burrower

Turtle Frog

African Bullfrog

Budgett's Frog

Helmeted Water Toad

Skull characteristics
Convergent characteristics in the skulls of small-prey dietary specialists, and large prey-feeding frogs. The Mexican Burrowing Toad (*Rhinophrynus dorsalis*), Guinea Snout-burrower (*Hemisus guineensis*), and Turtle Frog (*Myobatrachus gouldii*), are medium-sized burrowing frogs that feed on ants and termites; the African Bullfrog (*Pyxicephalus adspersus*), Budgett's Frog (*Lepidobatrachus laevis*), and Helmeted Water Toad (*Calyptocephalella gayi*) have heavily ossified skulls, and regularly consume much larger prey, from invertebrates to fish, frogs, birds, and small mammals.

PLANT-EATING FROGS

Perhaps the most extraordinary dietary specialism for frogs is found in those which intentionally include plants and plant material as a regular food source. Most, if not all frogs do ingest plant material, but this is because it is either inside their prey—in the stomach of a caterpillar, for example—or as a result of inadvertently capturing that material while feeding—pieces of vegetation gripped by the caterpillar as it is swallowed, or as by-catch sticking to the frog's tongue as it captures its meal.

↑ The Indian Green Frog (*Euphlyctis hexadactylus*) is unusual in that it complements its diet of invertebrates with aquatic plants.

↗ Izecksohn's Brazilian Treefrog (*Xenohyla truncata*) readily eats fruit, and will also enter flowers to feed on floral structures and nectar. As a result, it is also likely to be an important pollinator.

However, some species of frogs will regularly, and intentionally, feed on plant material. Perhaps one of the most extreme examples is the Indian Green Frog (*Euphlyctis hexadactylus*). Like many frogs, the tadpoles of this species are herbivorous, but once adult more than half their diet can consist of aquatic plants. Furthermore, these frogs may also be adapted for effectively grazing on plant material, as they have conical-shaped teeth—the same condition present in the teeth of many herbivorous mammals, and adapted for grasping and snipping off pieces of vegetation.

Including plant material in the diet also extends to frugivory. The Cururu Toad, for example, will take fallen fruits of the acerola cherry, while Izecksohn's Brazilian Treefrog (*Xenohyla truncata*, see page 184) has fruit as a significant part of its diet, to the extent that frogs will even compete for this important resource.

FROGS, FOOD, AND FEEDING

Ambush predators or active foragers?

↙ Some frogs, such as this Surinam Horned Frog (*Ceratophrys cornuta*), readily include smaller frogs in their diet.

↘ The European Treefrog (*Hyla arborea*) displays considerable agility, speed, and accuracy when capturing insect prey.

Whatever a frog's diet, there is always a trade-off between the energy expended finding food and the nutritional and energetic value of that food, with the ideal outcome being maximum benefit for minimum effort—the optimal foraging strategy.

Frogs are masters at conserving energy, only moving when necessary and then primarily to seek appropriate habitat and shelter, engage in reproductive activity, or to feed. Many species are ambush predators, but frogs are also active hunters and will even switch strategies to make the best of a feeding opportunity.

AMBUSH PREDATORS

Frogs that are primarily ambush (sit-and-wait) hunters face a theoretical dilemma. It may be a long time before a food item passes by, and it also needs to do so close enough to be captured and consumed. If the frog is more of a specialist, to get a meal it needs to be in the

right place at the right time. However, there is a trade-off between waiting for a long time for a meal to present itself and the value of that meal given the time it's taken to arrive. Therefore, when the next meal could be days away, it pays to be unselective and to take opportunities as they arise. This can mean both feeding on smaller prey and having the capacity to consume as large a prey item as possible.

Among frogs, perhaps the greatest proponents of this approach are members of the family Ceratophridae (the South American horned frogs, or wide-mouthed frogs). Often seen in the pet trade, these are charismatic anurans with a relatively small body, but a large head, and an especially large and wide mouth, hence they are also referred to as Pac-man frogs. Many ceratophrids are greater than 4 in (100 mm) in size—for example, the Surinam Horned Frog (*Ceratophrys cornuta*, see page 186)—and position themselves partially buried on the forest floor, with just the head showing, perfectly camouflaged against the leaf litter and forest debris. Here they wait for unsuspecting prey to pass by, and will take not only invertebrates but also lizards, snakes, other frogs, birds, and small mammals, which they swallow whole. However, these frogs will happily snack on whatever is available between larger meals and readily consume a host of small invertebrate prey such as ants.

ACTIVE FORAGERS

While the majority of frogs tend toward the sit-and-wait style of ambush predation, many employ an active search method for finding prey. One of the best-known groups to do this are poison frogs from the families Dendrobatidae and Mantellidae. Although actively searching for food increases the chances of an encounter with a predator and is best avoided, their diet—alkaloid-containing ants and mites—confers upon them a bioaccumulation of toxins, which the frogs readily advertise through their bright skin coloration and patterns, meaning that potential predators leave them well alone (see Chapter 7). Poison frogs that do not eat alkaloid-packed ants and mites tend to lack aposematic coloration and also employ a sit-and-wait feeding strategy, although an interesting adaptation is seen in the Imitating Poison Frog (*Ranitomeya imitator*, see page 218).

SWITCHING STRATEGY

Some frogs will switch from an ambush strategy to one of active foraging, and such examples have been observed in several families, notably the Microhylidae (narrow-mouthed frogs) and other burrowing species. A typical driver of this behavioral switch is associated with food availability and aggregations of prey. In the absence of food, actively foraging frogs will do so more widely, until they find an aggregation—a column of termites, for example. Then, active foraging will be replaced by a sit-and-eat (rather than sit-and-wait) mode of feeding, rather like humans driving across town to dine at a Japanese sushi-train (*kaitenzushi*).

FROGS, FOOD, AND FEEDING

Feeding mechanics

With many frogs having close to 360-degree vision, locating a potential meal does not even require a turn of the head, and providing it is within range, they are highly efficient at capturing prey. As such, frogs are principally visual hunters, but some are known to employ sound, smell, and touch to find a meal. Once prey is detected and within reach, most frogs use their tongues as sticky projectiles to capture and bring it to the mouth, sometimes also grabbing the item with their forelimbs to guide it in.

The responses of frogs to olfactory feeding cues are largely under-researched, but given the regularity with which some species will intentionally take non-moving or even dead food items, the importance of smell for frog ecology—and feeding in particular—may be much greater than currently understood. Similarly, auditory detection of prey is an area of anuran life history about which little is known, but examples include Cane Toads which, when feeding on Túngara Frogs (*Physalaemus pustulosus*), use the calls of the male frogs to home in on their prey. Similarly, Southern Toads (*Anaxyrus terrestris*) have been observed to respond to the sound of prey items moving through vegetation, alerting them to the presence of a potential meal.

← The Malaysian Narrowmouth Toad (*Kaloula pulchra*) is considered an ant specialist, but will also take other invertebrate prey.

MECHANISMS OF TONGUE PROTRACTION

There are three general modes of tongue deployment to capture prey: mechanical pulling, inertial elongation, and hydrostatic elongation. The first mode, mechanical pulling, is considered the ancestral condition of frog tongue protraction and is found in all tongued archaeobatrachian frogs and several neobatrachian families. The tongue in this condition is expansively attached to the floor of the mouth and tends to be short, rounded, and protractible by no more than 70 percent of the length of the jaw. As a result, when feeding, the frog will generally lunge toward its prey to bring it into contact with the tongue and thereby back to the mouth.

The second mode, inertial elongation, has evolved independently multiple times in anurans and is considered the most common means of tongue protraction. Here, the tongue is attached at the front of the mouth and also has a flap that can be projected as the appendage is driven forward. Although aiming the tongue is limited to a ballistic forward protraction, the benefits of inertial elongation mean that when feeding, a frog might not need to move, and thus the animal can rapidly flick the tongue in and out to feed, all the while remaining almost invisible to unsuspecting prey.

The third mode, hydrostatic elongation, is present in Africa's snouted-burrowers (genus *Hemisus*) and the narrow-mouthed microhylid frogs. This mechanism relies on muscular contractions that lengthen the tongue as its width reduces and confers an elevated accuracy while allowing the frog to protract its tongue either side of its head while facing forward. For a frog with a jaw length of about half an inch (10 mm), hydrostatic elongation can extend the tongue by 300 percent (to 1¼ in/30 mm).

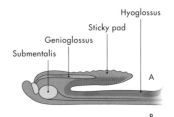

An extendable organ

From its resting position (A) and as the jaw is lowered, the tongue stiffens through contraction of the genioglossus and submentalis muscles, and then pivots over the submentalis (B, C) propelling the tongue forward, with its sticky pad directed toward the prey. Contraction of the hyoglossus muscle then pulls the tongue and prey into the mouth.

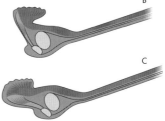

The muscular tongue

Protraction of a frogs' tongue is completed through inertial elongation, allowing them to flip out and fully extend this muscular organ, with the sticky end pad coming in to contact with a prey item (A, B) before returning it to the mouth to be swallowed (C).

Once a target has been located, the tongue is used to seize the prey and bring it toward the mouth. At this point, frogs will then often use their front feet to grab onto the food item and guide it in, upon which the very organs that have likely detected the prey can be brought into play once more, but this time to help the frog swallow its meal. The frog does this by retracting one or both eyes into the oral cavity, providing additional purchase on the food item to help push it into the gullet. While the tongue is the guiding mechanism, bringing the food to the mouth and toward the gullet, a combination of tongue, jaw, eye, and other functional movements aid in swallowing prey. Retraction of the eyes when feeding can also serve as a protective mechanism, especially when dealing with larger prey that could damage these critically important organs, and this combined with swallowing is why frogs will frequently blink and retract their eyes when feeding.

FROGS WITHOUT TONGUES

Some frogs, specifically members of the fully aquatic family Pipidae (comprising the genera *Pipa*, *Xenopus*, *Silurana*, *Hymenochirus*, and *Pseudhymenochirus*), do not possess a tongue, although they retain vestigial tongue musculature. The evolutionary loss of this characteristic frog appendage was likely driven by the resistance to operating such an organ in water (being around 800 times denser and about 50 times more viscous than air). Indeed, most other frogs that feed under water do not use their tongue as they do on land, and instead grab prey with their jaws.

Given the importance of the tongue in frog feeding, its loss therefore presents a challenge. In addition, pipids generally have reduced eyes and many are often found in turbid water, making visual location of prey difficult. To overcome these challenges, pipid frogs have evolved several physical and behavioral adaptations in their feeding ecology. Firstly, chemosensory detection, specifically smell, appears to play a significant role—ideal for detecting food in murky waters, and as pipids regularly scavenge, also for detecting inanimate food items.

Secondly, post-metamorphic pipid frogs retain the lateral line system they possess as tadpoles (see diagram on page 53), and this provides a tactile detection system similar to that of fish, in that once potential prey is within range the frogs are able to detect minute changes in water pressure caused by movement and/or the electric field given off by prey, although this is notably absent in the Congo Dwarf Clawed Frog (*Hymenochirus boettgeri*, see page 188) and the genus *Pseudhymenochirus*.

Thirdly, all pipid frogs employ inertial suction feeding, drawing water—and the food item with it—into the mouth (see diagram opposite). Finally, many pipid frogs use their front feet to scoop, push, grab, and otherwise manipulate food into the mouth. Notably, two of the larger pipid frogs, *Xenopus laevis* and the Surinam Toad (*Pipa pipa*, see page 126), have a grasping dexterity comparable to that of arboreal frogs, and *P. pipa* can even independently move its left and right mandibles to manipulate a meal.

FEEDING MECHANICS

INERTIAL SUCTION IN AQUATIC FROGS

Inertial suction is not only a feeding mechanism employed by frogs, but also the most common prey-capture technique found in aquatic vertebrates, including most fish. The process of engulfing a food item is driven by a rapid expansion of the buccopharyngeal cavity, which generates a temporary decrease in pressure within. This process causes water to flow toward the area of reduced pressure, carrying the food with it, ultimately to be swallowed by the frog.

Dwarf clawed frogs from the genera *Hymenochirus* and *Pseudhymenochirus* engulf prey using only inertial suction, and unlike other pipids do not employ their forelimbs to manipulate food. This mode of feeding in frogs was previously thought only present in pipids (members of the oldest frog group Archaeobatrachia). However, inertial suction feeding has since been seen in neobatrachian frogs, namely one species of the largely aquatic/semi-aquatic family Telmatobiidae. The Laguna de los Pozuelos' Rusted Frog (*Telmatobius rubigo*)—a fully aquatic species—was given mealworms in the lab and its feeding behavior filmed using high-speed cameras. This revealed remarkable similarities with the feeding behavior of pipid frogs, with *T. rubigo* lunging toward the prey, followed by the mouth opening in close proximity to the mealworm and prey being drawn into the buccopharyngeal cavity.

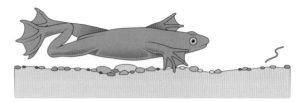
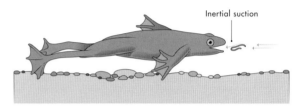
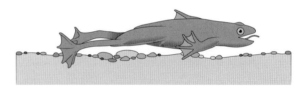
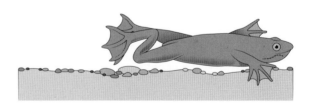

← With few exceptions, adult frogs swallow their prey whole, whether it's a spider (as here with this Red-eyed Treefrog, *Agalychnis callidryas*) or a mouse. The eyes of a frog do more than simply spot prey; they help facilitate the passage of a food item toward the gullet to be swallowed.

Prey capture in pipid frogs

After getting within striking range of their prey, pipid frogs snap their mouth open, creating a sudden force that draws the food into the mouth. Using this feeding mechanism—known as inertial suction—a frog can engulf its meal in less than 0.2 of a second, which roughly equates in human terms to the blink of an eye.

FROGS, FOOD, AND FEEDING

RHINOPHRYNUS DORSALIS

Mexican Burrowing Toad

An ancient lineage that is today the sole member of its family

SCIENTIFIC NAME:	*Rhinophrynus dorsalis*
FAMILY:	Rhinophrynidae
LENGTH:	2–3½ in (50–85 mm)
LIFE HISTORY:	Aquatic eggs (from hundreds to several thousand singly and in small clumps) and tadpoles; terrestrial (subterranean) adults
NOTABLE FEATURE:	The only frog adapted to feed underground
IUCN RED LIST:	Least Concern

This distinctive anuran has a rotund body, no neck, a small head, and a snout covered with sensory tubercles. A generally dark background coloration s dotted with contrasting lighter blotches and a dorsal stripe. Spadelike structures on its hind feet are used to tunnel backward through relatively soft soils in habitats adjacent to coastal areas from the southern ip of Texas through Mexico to Central America. The Mexican Burrowing Toad is almost entirely subterranean, appearing above ground only when heavy rains fill temporary pools and stimulate them to emerge and breed, which can happen as infrequently as every two years.

The Mexican Burrowing Toad has a mode of feeding that is not only different from other frogs that feed on insects but is unique among anurans. Unlike other burrowing frogs, which may gorge themselves on insects that appear on the surface during short rainy periods, this species feeds only on ants and termites underground. To do this, it has a reinforced skull, a short stiffened jaw, and a robust snout that it can push into and through the hardened walls of insect burrows. Whereas other frogs open their mouths wide and flick out their tongues to catch prey, the Mexican Burrowing Toad has a completely different muscular mechanism controlling its tongue, which can be protruded rigidly through a barely-opened mouth. Only the tip of the tongue needs to be extruded, an adaptation that likely helps the frog secure its ant and termite prey in their narrow underground tunnels. The roof of the mouth and gullet also have complex ridges and folds, which may allow the frogs to withstand the ferocious bites of ants and termites as they are caught and swallowed. This unique combination of features means that the Mexican Burrowing Toad is highly distinctive in evolutionary terms, and so is placed in a family all of its own.

The breeding period for Mexican Burrowing Toads lasts just a few days, with many individuals emerging from their subterranean hideaways following May rains. They breed in temporary pools, where females may lay several thousand eggs—the number of eggs ranges from around 600 to nearly 8,000, depending on the size of the female (larger frogs can produce a greater number of eggs). Tadpoles hatch and complete their metamorphosis into juvenile frogs in a few weeks, and sexual maturity is reached in 8–9 months.

→ The flaccid body and loose skin of the Mexican Burrowing Toad is an adaptation to an almost entirely subterranean lifestyle. These anatomical adaptations allow it to easily move through the soft soil and narrow burrows. This outward appearance belies a highly robust skeleton, the skull in particular.

FROGS, FOOD, AND FEEDING

LEPIDOBATRACHUS LAEVIS

Budgett's Frog

Cannibalistic carnivore

SCIENTIFIC NAME:	*Lepidobatrachus laevis*
FAMILY:	Ceratophryidae
LENGTH:	2–7 in (50–120 mm)
LIFE HISTORY:	Aquatic eggs (up to 1,400), tadpoles, and adults
NOTABLE FEATURE:	Large flat-bodied frog with a robust head that accounts for a third of its total size
IUCN RED LIST:	Least Concern

This dark-green to grayish frog lives and breeds in shallow ponds in the semiarid lowland forest areas of Paraguay, Bolivia, and Argentina. Its bizarre appearance has made it popular within the pet trade. Budgett's Frog survives the dry winter months buried in mud within a waterproof cocoon consisting of several layers of dead skin. Adults emerge when the rains return and refill the many shallow pools of the Gran Chaco region of South America.

Budgett's Frogs are pugnacious feeders and predators. Their flattened body allows them to sit motionless in the water with only the eyes protruding above the surface. Similarly to the aquatic pipid frogs, Budgett's Frog possesses lateral line organs in the skin that can detect the movements of potential prey (and predators) in the water. Should an insect, snail, or other frog come within range, a Budgett's Frog lunges forward to capture it in their enormous jaws. Large teeth on the upper jaw and two fanglike projections on the lower jaw help subdue struggling prey.

A member of the family Ceratophryidae—known for their propensity to consume very large prey (often of nearly equal size) whole—Budgett's Frog stands out among even this voracious family. Ordinarily, anuran tadpoles have a decidedly larval internal morphology and skeletal structure, and they are adapted to grazing and filtering small particles of food.

It is during metamorphosis into the adult form that they undergo almost complete remodelling. Budgett's Frogs are different: during embryonic development, their tadpoles essentially acquire both an adult stomach and an adult jaw structure, which allows them to subdue and swallow very large active prey items whole almost immediately after leaving the egg. This physiology and behavior therefore continues unbroken throughout the life cycle of the animal—from tadpole to frog—and is unique among anurans.

Despite being aquatic, Budgett's Frog is a relatively poor swimmer, but it is highly adapted to the seasonal ephemeral pools in which it lives, exemplified by the fact that tadpoles may reach metamorphosis in as few as 20 days, and be able to breed in less than a year. When threatened, a Budgett's Frog will inflate its body on stiffened limbs. If this fails to deter a would-be attacker, the frog may lunge toward the threat with its mouth open, and even produce a high-pitched scream. This has led to the Guarani name for the species of kukurú-chiní, which translates as "the toad that shrieks."

→ Budgett's Frog is an aquatic sit-and-wait predator. With nostrils and eyes positioned on top of its head, much like a crocodile it will sit motionless in the water waiting for a meal to pass within reach.

XENOHYLA TRUNCATA

Izecksohn's Brazilian Treefrog

The nectar, fruit, and flower-eating frog

SCIENTIFIC NAME:	*Xenohyla truncata*
FAMILY:	Hylidae
LENGTH	1¼–1¾ in (30–45 mm)
LIFE HISTORY	Not well known; aquatic tadpoles, terrestrial adults
NOTABLE FEATURE	This frog may be the only known amphibian seed disperser and pollinator
IUCN RED LIST	Near Threatened

Known only from the coastal lowlands of Rio de Janeiro, Brazil, this unassuming, yellow-brown treefrog spends much of its time hiding in terrestrial bromeliads, but is remarkable for its nocturnal foraging habits.

Although Izecksohn's Brazilian Treefrog also includes a range of arthropods in its diet, it snacks on fruits and petals, and also nectar, which it consumes with suctionlike movements. Fruits of several native plant species contribute to its diet (*Maytenus obtusifolia*, *Erythroxylum ovalifolium*, *Anthurium harrisii*, and *Cordia taguahyensis*), and the frog has even been observed feeding on the flowers of an introduced alien plant, the Bearded Iris. The fruit-eating habits of the frogs are dictated by plant phenology, with different fruits available at different times of year, thereby making the fruiting bodies of its favored plants a changeable but continual food resource upon which these frogs rely.

While knowledge of the full relationship with its plant prey is incomplete, Izecksohn's Brazilian Treefrog is known to be a seed disperser. This ecological role—common in birds and mammals—was verified following the collection of several frogs, and the observation that they apparently defecated seeds back in the laboratory. Further investigation found seeds in both the stomach and intestine of the frogs, confirming their status as frugivorous seed dispersers. While tadpoles of many frog species act as seed dispersers during their aquatic (and often entirely herbivorous) life stage, to date this is the only known occurrence of seed dispersal in a post metamorphic amphibian.

Due to their propensity for feeding on flowers, and also entering them to suck up nectar, these frogs regularly exit flowering bodies with pollen grains attached to their skin. This pollen is therefore ready for transfer to a new flower, and—in another amphibian first—may mean that Izecksohn's Brazilian Treefrog is not just the only known amphibian seed disperser, but also the only known amphibian pollinator.

→ While fruit feeding by Izecksohn's Brazilian Treefrog was first reported in the late 1980s, its taste for nectar was unknown until 2020. It is possible that nectar feeding may be especially important during the breeding period, which is an energetically costly time for both male and female frogs.

FROGS, FOOD, AND FEEDING

CERATOPHRYS CORNUTA

Surinam Horned Frog

Cryptic ambush predator

SCIENTIFIC NAME:	*Ceratophrys cornuta*
FAMILY:	Ceratophryidae
LENGTH:	2¾–4¾ in (72–120 mm)
LIFE HISTORY:	Aquatic eggs (500–2,270 in clumps) and tadpoles; terrestrial adults
NOTABLE FEATURE:	The mouth is 1.6 times wider than the length of its body
IUCN RED LIST:	Least Concern

The Surinam Horned Frog is a dedicated, sit-and-wait hunting strategist, often remaining motionless night after night in its quest for a meal. There is not much that isn't on the menu, and these large frogs will readily consume prey across a range of sizes, from ants and beetles, to reptiles, small mammals, and birds. Their diet also includes other frogs—even members of their own species.

Often encountered buried in leaf litter, with only its head visible, the Surinam Horned Frog has a highly variable cryptic coloration of greens and browns combined with shape-diffusing markings. This coloration provides perfect camouflage on forest floors across the Amazon basin, making the Surinam Horned Frog a master ambush predator.

Somewhat poetically, a regular anuran prey item in the diet of these wide-mouthed frogs is one of the narrow-mouthed frogs (Microhylidae), the Bolivian Bleating Frog (*Hamptophryne boliviana*). Like the adults, the tadpoles of the Surinam Horned Frog are carnivorous, and those of other frog species are a major component of its diet.

The Surinam Horned Frog has a large, bony (hyperossified) skull and fang-like teeth, including two large fangs projecting from the lower jaw. These are not, in fact, teeth but rather upward projections of bone, all features shared with other members of the Ceratophryidae. The tendency for larger, more robust bone to fossilize well means the extent of skeletal ossification in ceratophrid frogs, in contrast to other anurans, has likely contributed to their relatively rich fossil record.

→ The Surinam Horned Frogs rely on camouflage and concealment to catch their prey. They bury the posterior and lower section of their bodies in leaf-litter by moving it aside with their back legs, leaving the head and front legs emergent but perfectly blended into the surroundings, and ready to take the next opportunity for a meal.

FROGS, FOOD, AND FEEDING

HYMENOCHIRUS BOETTGERI

Congo Dwarf Clawed Frog

A fully aquatic frog, common in the aquarium pet trade

SCIENTIFIC NAME:	*Hymenochirus boettgeri*
FAMILY:	Pipidae
LENGTH:	1–1½ in (25–35 mm)
LIFE HISTORY:	Aquatic eggs (around 450), tadpoles, and adults
NOTABLE FEATURE:	These frogs have no tongue and feed using suction to engulf prey
IUCN RED LIST:	Least Concern

The Congo Dwarf Clawed Frog is found across Africa's Congo basin and equatorial forest from southern Nigeria and Gabon to the eastern Democratic Republic of Congo. Despite being fully aquatic, this species (and other *Hymenochirus* frogs) lacks the lateral line system for detecting movement in the water which is present in other pipid frogs. It prefers habitat with shade, where its brown-gray, dappled coloration matches the substrate of the lentic (still) or slow-flowing water bodies in which it lives. While both tadpoles and adults are carnivorous and employ suction feeding, their methods of predation are quite different.

Despite their small size, both tadpoles and adults have a proportionally large buccal (oral) cavity, which they use for suction feeding, but the feeding mechanism of tadpole and frog differs considerably. A few days after hatching, tadpoles of the Congo Dwarf Clawed Frog become dedicated hunters, actively pursuing small invertebrate prey. What makes the tadpoles so remarkable is that they not only feed in a similar way to many fishes, but are more effective at doing so than similarly sized fish. Once a prey item has been chased down, the tadpoles extend their rounded, tubular mouth while expanding the buccal cavity, thereby sucking in the target through the oral tube for swallowing whole. Conversely, the metamorphosed frogs are considered much less active hunters and can be observed motionless but in a "ready position," waiting for a prey item to come within range.

The skin secretions of the Congo Dwarf Clawed Frog have been found to contain a varied source of antimicrobial peptides (short strings of chemicals also related to building proteins), and so named hymenochirins. These peptides can inhibit the growth of several bacteria and have the potential for clinical application to fight cancers and viruses.

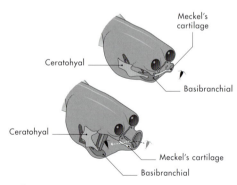

Feeding apparatus
Congo Dwarf Clawed Frog tadpoles use specialized morphology, together with their extendible tubelike mouth, to ingest prey.

→ Like other members of the family Pipidae, the Congo Dwarf Clawed Frog is tongueless, and feeds instead using inertial suction (see page 179).

DEFENSES AGAINST PREDATORS & PATHOGENS

DEFENSES AGAINST PREDATORS AND PATHOGENS

The challenge of predation

The eggs, tadpoles, and post-metamorphic stages of frogs provide food for a host of predators, from invertebrates to fishes, reptiles, birds, mammals, and even other frogs. With a relatively soft, thin skin, their bodies can be easy to pierce, bite, swallow, and digest, and many frogs congregate in large numbers when breeding and/or hibernating, presenting an easy meal. They also lack the powers of escape—running or flying—seen in mammals and birds.

As such, frogs have an important role to play in food chains, and indeed in the lives of the species that eat them. Nevertheless, frogs are not without defense and have evolved some of the most sophisticated systems in the animal kingdom to protect themselves against predators and pathogens.

Unlike reptiles, mammals, and birds, frogs lack a waterproof skin. Consequently, they live in damp places, see out dry periods underground, or conserve water in other ways (see Chapter 2). As frogs breathe through their skin it must remain soft and moist, which means it cannot be molded into mechanical devices such as scales, hair, and feathers for protection. However, the frog skin serves other important purposes, containing a battery of color pigments, chemicals, and microorganisms that are crucial to frog survival.

← The Amazonian Milk Frog (*Trachycephalus resinifictrix*), whose skin texture and coloration resembles a bird dropping, owes its name to the milky secretion exuded from its skin when handled.

↗ The Australian Green Treefrog (*Litoria caerulea*)—also known as the Dumpy Treefrog or White's Treefrog—secretes a fluid from its skin that has antimicrobial properties.

THE CHALLENGE OF PREDATION

HOW FROGS RESPOND

When confronted by a predator, a frog's initial response may be either to escape as quickly as possible by rapid jumping or swimming, or to remain motionless and rely on camouflage to hide from the assailant. Although two very different responses, which one will prove most effective depends on the sensory capabilities of the predator, the defense capabilities of the frog, and the environmental circumstances of the attack. Indeed, if remaining motionless initially fails to deter the predator, the same frog may then resort to leaping away. Natural selection results in genes that have enabled frogs to survive predation being passed on to the next generation. For example, individual frogs that are brightly colored, indicating that they may be toxic and not good to eat, are more likely to survive and pass on those good "bright color" genes to their offspring. In contrast, frogs of the same species whose coloration genes make them, by chance, dull-colored are more likely to be eaten and less likely to pass on dull color genes to their offspring. However, the same evolutionary processes are driving adaptation in the predators of frogs, and those that have genes for, say, good color vision and toxin tolerance may be able to capture more frogs than those that don't. Consequently, there is an evolutionary "arms race" in nature, with improved predation mechanisms driving ever-more sophisticated ways of avoiding predation.

DEFENSES AGAINST PREDATORS AND PATHOGENS

THE FIRST LINE OF DEFENSE

The outer layer of cells in a frog's skin is called the stratum corneum which contains keratin. This layer of skin is periodically shed and often peeled off in one piece after breaking around the mouth. Below this lies the epidermis and then the deeper dermis. The dermis also contains numerous glands with short tubes to the surface of the skin that serve different functions. Mucus glands keep the skin moist, while granular glands secrete an array of chemicals that are used for defensive purposes. The granular glands may be aggregated into larger, wart-like structures, such as the conspicuous parotoid glands on either side of the heads on toads. These may squirt noxious substances at predators if the animal is threatened.

Also lying within the dermis are three types of chromatophore, which determine the color of the frog. Melanophores contain the dark pigment melanin. When the melanin is concentrated in the center of the cell, the skin will appear light in color. However, melanophores have many tiny channels that reach into the spaces between other cells, and when the melanin is dispersed along these channels the skin darkens.

Xanthophores lie between the epidermis and the dermis and contain yellow, orange, or red pigments. Below the xanthophores lie iridophores. Instead of containing pigments, these cells reflect light of different wavelengths through the skin and thereby control the brightness of the colors in the xanthophores. In the absence of xanthophores the iridophores may reflect blue light and impart a blue hue to the skin.

The interactions between these three types of cells are regulated by environmental conditions and determine whether the skin appears dark or light, and brightly colored or well-camouflaged. In addition, a frog's skin contains a complex array of chemicals that inhibit disease-causing bacteria and fungi.

→ Bright red patches against the contrasting black background of the Koe-Koe (*Oophaga solanensise*) warn predators of its toxicity.

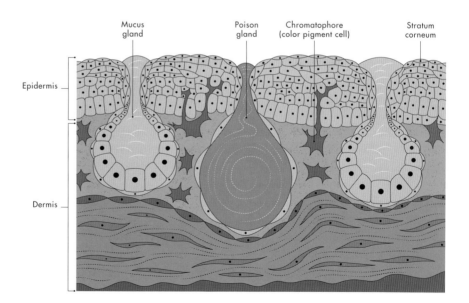

Frog's skin
Cross section of the skin of a frog showing the secretory glands and chromatophores.

DEFENSES AGAINST PREDATORS AND PATHOGENS

Tadpoles: anti-predator behaviors

Birds, snakes, fishes, beetles, and dragonflies are just some of the predators that will make a meal of tadpoles. Although the production of large clutches of eggs and many resultant tadpoles which can then form dense aggregations confers increased individual survivability in the event of a predator attack (see Chapter 2), tadpoles have a variety of strategies for reducing the risk of being eaten.

BEHAVIORAL AND DEVELOPMENTAL CHANGES

Many species of frogs—including Natterjack Toads (*Epidalea calamita*, see page 246) in Europe and Wood Frogs (*Lithobates sylvaticus*, see page 72) in North America—can avoid water bodies that contain certain predators and competitors. If eggs are laid in pools containing predators, reducing swimming may mean the tadpoles will be overlooked by a fish that might otherwise detect any movement in the water.

Speeding up development means that the tadpole can metamorphose faster and escape a predator-rich water body. Although growth and development are linked, slower growth and faster development results in the tadpole metamorphosing quickly but into a smaller froglet. But this comes at a price, as smaller froglets may not be able to jump as far as larger froglets and so could also be more vulnerable to predation.

← While still in the egg, Red-eyed Treefrog (*Agalychnis callidryas*) tadpoles can detect when a predator begins to feed on the clutch, and will hatch early to avoid being eaten.

→ In the presence of dragonfly predators, Pine Woods Treefrog (*Dryophytes femoralis*) tadpoles develop deeper, colored tailfins to divert attacks away from more vulnerable parts of the body.

TADPOLES: ANTI-PREDATOR BEHAVIORS

Laurel and Hardy effect
The "Laurel and Hardy" effect as seen in tadpoles of the Mallorcan Midwife Toad. The streamlined, snake-escaping swimmer "Laurel" and the more robust "Hardy" when snakes are absent from natal ponds.

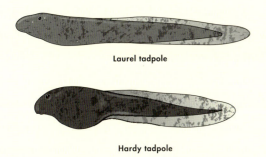

Laurel tadpole

Hardy tadpole

Developing tadpoles of the Red-eyed Treefrog (*Agalychnis callidryas*, see page 210) can even speed up hatching if a predator starts to eat the eggs.

Flexibility in hatching and tadpole development occurs in many frog species and manifests itself in different ways. In the presence of aquatic predators, the tadpoles of many species will spend more time in vegetation or reduce their swimming activity in exposed areas. Such behavioral changes are also linked to growth, size, and even shape. Tadpoles of the Mallorcan Midwife Toad (*Alytes muletensis*, see page 102), for example, reduce their swimming activity in the presence of the predatory Viperine Snake (*Natrix maura*).

ALTERED APPEARANCE

In addition to changing behavior, the presence of predators can cause tadpoles of a range of species to change shape, size, or color to reduce the risk of being eaten. Again, in the presence of predators Mallorcan Midwife Toad tadpoles are streamlined with thick tail muscles, so they can escape snakes with a rapid undulation of the tail. However, when snakes are absent the tadpoles have wide heads and mouths better equipped for feeding. This change in shape according to predator threats has been termed the "Laurel and Hardy effect". Such "inducible defenses," which can develop in as little as a couple of weeks, may be advantageous in habitats where the presence or absence of predators is variable and tadpoles need to change their body form according to the prevailing risks (see the Southern Gray Treefrog, *Dryophytes chrysoscelis*, page 212).

SEMITERRESTRIAL TADPOLES

Although tadpoles are often regarded as the aquatic stage in the frog life cycle, many species produce eggs and tadpoles on land (see Chapter 2). Indeed, the tadpoles of frogs in at least six different families live on flat, wet rocks adjacent to streams. When threatened or disturbed, these tadpoles can jump. When attacked by ants, for example, tadpoles of the Rock Frog (*Thoropa taophora*) from Brazil can flick their tail and leap into the air to escape. Species with such semiterrestrial jumping tadpoles are characterized by long, thin tails with much-reduced fins.

DEFENSES AGAINST PREDATORS AND PATHOGENS

Adults: anti-predator behaviors

As well as escaping by swimming or jumping to evade a predator, frogs may also adopt a range of different postures to avoid becoming a meal. If remaining motionless and relying on camouflage fails, they may simply "play dead" (sometimes called "death-feigning," or thanatosis), lying on their backs with their legs limply protruding from their sides.

↑ The Yellow-bellied Toad (*Bombina variegata*) from Europe is well-camouflaged on top, but if threatened will bend its body into the "unken reflex," displaying the bright yellow warning coloration on the underside.

BODY POSTURE

Death-feigning has so far been observed in over 40 species and across a range of families but is more often seen in frogs lacking toxic skins. In the Central African Spiny Frog (*Acanthixalus spinosus*), the tongue is extruded while playing dead. Alternatively, a motionless frog may close its eyes and curl its legs into the body. This type of behavior, known as "shrinking" or "contracting," has been observed in the Bahia Green Treefrog (*Phyllomedusa bahiana*) from Brazil and occurs more frequently in those frogs with toxic skins. Many predators, including birds and snakes, are stimulated by moving prey and will ignore motionless frogs. Likewise, unless the predator is a specialist carrion-feeder it will probably ignore prey that it perceives as dead.

In addition to remaining motionless, frogs may adopt one or more different postures to deter predators (see the diagram opposite). Some bufonid toads will blow up like a balloon by inflating their lungs with air to make their body appear much larger, particularly if threatened by predators such as snakes that would otherwise swallow the animal whole. A similar impression may be made by arching the body on stretched back legs. Some frogs, such as Perez's Snouted Frog (*Edalorhina perezi*) from South America, arch their back in this way to expose two large, black eyespots, which may serve to startle a would-be attacker.

Arching the body in the opposite direction, so that all four legs are off the ground with the head pointing upward, is known as the "unken reflex." This exposes a brightly colored throat and belly to the predator, warning that its potential meal is toxic (see Be Warned, page 208). Such behavior occurs in a variety of frog families but is particularly well known in the fire-bellied toads (genus *Bombina*) from Europe and Asia. These toads are cryptically colored on the upper surface but have bright yellow or red undersides. Interestingly, the bright pigments in the belly take time to develop, and young toads that lack them do not display the unken reflex.

OTHER DEFENSES

Several species of frogs, including the European Common Frog (*Rana temporaria*, see page 32), have been reported to emit a loud, scream-like distress call when attacked or injured. This is produced with the mouth open. Although the purpose of such a call is unclear, mouth-gaping may expose the light-colored lining of the mouth which may contrast with the frog's background color, with the aim of startling the attacker.

Although frogs have blunt teeth, which are unlikely to break skin, some species, such as horned frogs in the South American genus *Ceratophrys*, will defend themselves by attempting to bite a would-be attacker, including humans. One of the more unusual defensive behaviors is seen in the Pebble Toad (*Oreophrynella nigra*), known from rocky, mountainous habitats in Venezuela. When threatened, the toad tucks in its legs, makes its body rigid, and simply rolls away, tumbling and bouncing as it progresses down the rocky slope.

On defense
Examples of defensive postures used by frogs when threatened.

Crouching and arching the back

Arching the body to expose warning colors (unken reflex)

Mouth gaping

Raising the body to expose eye spots

Death feigning (thanatosis) and tongue protrusion

DEFENSES AGAINST PREDATORS AND PATHOGENS

Chemical and biological defenses

All life history stages of frogs possess chemical defenses to some extent, although these vary between species and may change as predation pressures shift while frogs develop and grow. In addition, the skin of frogs contains chemicals and microorganisms that can inhibit the disease-causing pathogens.

IN EGGS AND TADPOLES

Although egg clutches often suffer high mortality, those of some species are unpalatable to certain predators. In North America, for example, the eggs of both the Green Frog (*Lithobates clamitans*, see page 160) and American Bullfrog (*L. catesbeianus*, see page 96) tend to be distasteful to newts and salamanders. Chemical defenses that confer protection on eggs are sometimes carried over to the tadpole stage. European Common Toads (*Bufo bufo*, see page 104), for instance, can breed in permanent lakes and ponds largely because their eggs and tadpoles are distasteful and avoided by would-be fish predators. In contrast to largely palatable ranid tadpoles that are cryptically colored, bufonid tadpoles are conspicuously black in color and this may serve as a warning of their potential unpalatability.

↗ The Mantella frogs have convergently evolved similar natural histories and warning coloration patterns to the unrelated poison frogs of South America.

→ The Strawberry Poison Frog (*Oophaga pumilio*) has up to 30 different color forms in different parts of its range.

← Some frogs, such as the American Bullfrog (*Lithobates catesbeianus*), lay eggs which are distasteful to many predators. However, there is little protection if water levels fall, and eggs are left high and dry.

CHEMICAL AND BIOLOGICAL DEFENSES

IN ADULTS

It is in adult frogs that the most formidable array of chemical defenses has evolved, and these have been particularly well-studied in several families, collectively known as the "poison frogs." Comprising species found across both the Eastern and Western hemispheres, frogs from the families Dendrobatidae, Bufonidae, Mantellidae, Eleutherodactylidae, and Myobatrachidae demonstrate how similar mechanisms of chemical defense have evolved convergently in anurans in different parts of the world. Although not closely related, the brightly colored *Mantella* frogs of Madagascar share many biological features with the poison frogs of Central and South America.

However, there is enormous variation in the range of chemicals in the skin used by frogs for defense, both within and between species.

For example, there are currently at least 1,400 different "lipophilic alkaloids," which are known to occur in frog skin, and this is just one group of chemicals deployed by frogs for defense purposes. The Strawberry Poison Frog (*Oophaga pumilio*, see page 128) shows huge variation in color patterns between the islands off the coast of Panama where it occurs and, intriguingly, the different populations have different types of alkaloid chemical defenses in their skin.

One reason for such extensive variation is that chemicals are sequestered from the frogs' diet. Ants, mites, beetles, and millipedes all contain alkaloids and are eaten by frogs, which then retain these chemicals for storage in their skin glands. As a result of their attractive colors, poison frogs are popular in the pet trade, but captive-bred frogs lack the alkaloids found in wild frogs because they are usually fed an alkaloid-free diet. Likewise, the fact that the alkaloid-containing arthropods eaten by the frogs may vary seasonally and between locations can also contribute to the variation seen in chemical defenses. The Australian poison frogs in the genus *Pseudophryne* are unusual in that they can synthesize their own alkaloids as well as using those from their prey.

The skin toxins of three species of dendrobatid frogs have been used to tip the blow-pipe darts used by Indigenous hunters in Columbia: the poison dart frogs *Phyllobates terribilis*, *P. aurotaenia*, and *P. bicolor*. Their skins contain the alkaloids batrachotoxin, homobatrachotoxin, and batrachotoxinin A, with the Golden Poison Dart Frog (*P. terribilis*) renowned as one of the most poisonous animals in the world—just one milligram of its skin toxin is considered enough to kill ten people. Although such secretions usually have to enter the bloodstream to be effective, great care is needed when handling wild Golden Poison Dart Frogs to prevent their toxins entering the body. Unsurprisingly, these frogs have relatively few predators, although one species of snake—*Erythrolamprus epinephalus*—can apparently consume juvenile Golden Poison Dart Frogs, as well as other toxic frogs, without suffering ill effects.

The skin defenses of poison frogs therefore result in predators finding them so distasteful that they learn to avoid them in their diet. As their toxins need to be ingested in some way to take effect, frogs are considered poisonous. However, two species of Brazilian frogs can inject their poison and are truly venomous.

BRAZIL'S VENOMOUS FROGS

Venomous animals are those that have developed a mechanism for delivering poison into a would-be attacker. Two species of Brazilian hylid frogs have achieved this. Greening's Frog (*Corythomantis greeningi*) and Bruno's Casque-headed Frog (*Aparasphenodon brunoi*) have sharp, bony spines on the head which emerge through the toxin-containing granular glands in the skin. Unlike other frogs, both species have unusually mobile heads and when threatened jab an assailant by actively flexing the head up and down or from side to side. In this way, the frogs inject their poison directly into the bloodstream of the attacker. Both species produce venom that can cause intense pain for several hours in humans and is more toxic than the local pit vipers (genus *Bothrops*).

Although the toxicity of Greening's Frog is less than that of Bruno's Casque-headed Frog, it has more head spines and larger poison glands, so it can potentially inject a greater volume of venom. Both frogs are therefore truly venomous rather than poisonous.

↓ Bruno's Casque-headed Frog (*Aparasphenodon brunoi*) has a highly ossified, wedge-shaped skull covered in toxin-delivering spines, making this species one of only two venomous frogs.

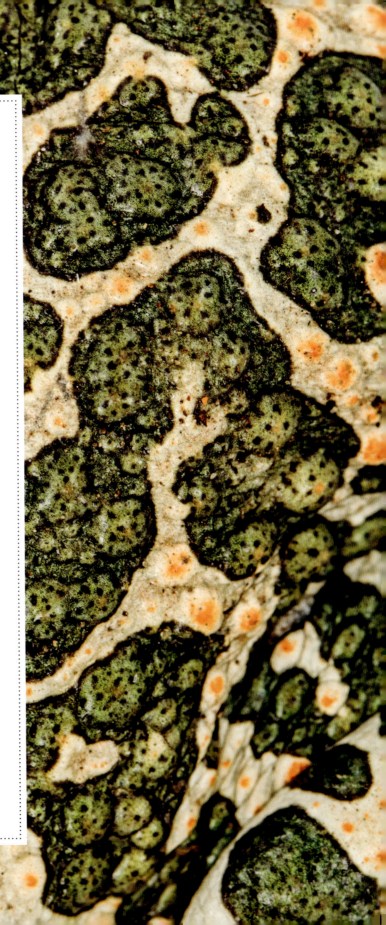

FROG SKIN: ANTIMICROBIAL PEPTIDES AND BACTERIA

In addition to predators, frogs are also threatened by infections from various pathogens that cause disease and mortality. One of these, the chytrid fungus *Batrachochytrium dendrobatidis* (*Bd*), is responsible for the decline and extinction of frogs on a global scale (see Chapter 9). Recent research has shown that some frogs have two lines of defense in their skin which can inhibit the fungus. Firstly, they produce fungi-inhibiting secretions from skin glands and, secondly, the skin supports an assemblage of fungi-inhibiting bacteria. More than a thousand different chemicals—antimicrobial peptides—have been identified from frog skin in recent years. Different frogs have different cocktails of peptides, which facilitates protection across a range of potential pathogens. These peptides also contribute to defense against potential predators.

Due to the way antimicrobial peptides destroy the cells of a pathogen, the pathogen is unable to develop any resistance over time. As such, the antimicrobial peptides from frog skin have the potential to provide a new generation of resistance-proof antibiotics for a wide range of diseases in both humans and other animals. Fungi-inhibiting bacteria and the microbial communities that naturally occur on frog skin—the skin microbiome—also come in a diversity of forms able to mop up potentially infectious fungi, diseases, and parasites.

The presence or absence of antimicrobial peptides and functioning skin-microbiome may explain why some frogs are much more susceptible to disease than others. There remains the intriguing possibility that some peptides may destroy pathogenic bacteria while benefitting those bacteria that can inhibit the fungus. Either way, ongoing research on both may eventually reveal a mechanism for controlling the disease that is killing frogs across the world.

→ Close-up of the skin of the Green Toad (*Bufotes viridis*) showing the pores on the glands in the skin through which secretions are released.

DEFENSES AGAINST PREDATORS AND PATHOGENS

Camouflage, warning coloration, and mimicry

The color patterns of frogs serve to conceal them from predators or advertise that they are toxic. Some frogs are also able to mimic the warning coloration of other species in order to enhance their chances of survival.

CAMOUFLAGE, WARNING COLORATION, AND MIMICRY

→ Dark bands on the hind legs of the North American Leopard Frog (*Lithobates pipiens*) help to break up its outline. This type of patterning is present in many anurans.

↙ The Vietnamese Mossy Frog (*Theloderma corticale*) is almost invisible when at rest in its favored habitat.

Tadpoles display an astonishing diversity of form and function (see Chapter 2) but tend not to show extensive color variation, instead being cryptically patterned in shades of gray, green, or brown to match their habitat. Some tadpoles—usually those that are toxic and unpalatable—are conspicuously black and often form large aggregations as a defense against predators. Although their skins contain chromatophores (see page 194), color change in tadpoles is largely confined to a lightening or darkening in relation to environmental conditions.

CAMOUFLAGE

In contrast, post-metamorphic frogs have a range of defenses, from color patterns that make them difficult to distinguish from the surface on which they are resting to bright colors that conspicuously communicate to predators that they are toxic to eat. Camouflage involves matching the colors to the background, breaking up the outline of the frog so it is difficult to see and masking conspicuous characteristics such as eyes and legs (for example, see the Long-nosed Horned Frog, *Pelobatrachus nasutus*, page 214).

Nearly all frogs can vary how light or dark their skin color is, usually in relation to prevailing light and temperature conditions (although advertising their presence to females or rival males may also play a role—see Chapter 3).

Breaking up the body outline may involve developing unusually shaped appendages or having contrasting bands of light or dark color that resemble shadows. The eye of a frog is something a predator could easily fixate upon, so this is often masked by a black band, as in many ranid frogs. When the hind legs are folded against the body, as in a frog at rest, dark bands running across the legs may further help disguise the outline of the animal. The most well-camouflaged frogs display a combination of these features, but the Mutable Rainfrog (*Pristimantis mutabilis*, see page 216) takes the art of camouflage one step further by varying the surface texture of its skin.

CAMOUFLAGE, WARNING COLORATION, AND MIMICRY

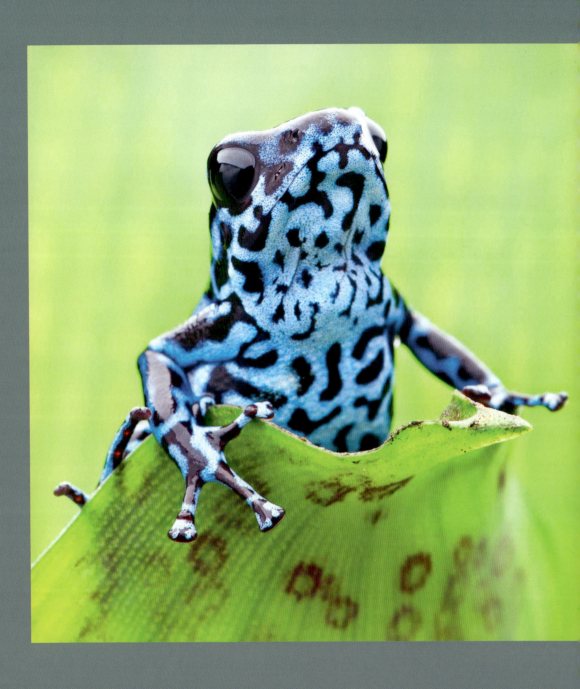

↖ The Long-nosed Horned Frog (*Pelobatrachus nasutus*) is incredibly difficult to spot in the leaf litter of its forest floor home.

↑ Despite its common name, the Strawberry Poison Frog (*Oophaga pumilio*) has many color morphs. Active during the day, this frog's bright coloration and patterning advertise its toxicity to would-be predators.

DEFENSES AGAINST PREDATORS AND PATHOGENS

BE WARNED

Conversely, it may be advantageous for a frog to advertise to predators that it is not good to eat. Consequently, frogs that contain toxins in the skin—particularly the poison frogs—are often bright shades of red, yellow, green, and blue, which strongly contrast with often present darker patterning, and against the background features of their habitat. Using coloration to advertise that you may be unpalatable or noxious is known as aposematism, and widespread in the animal kingdom (for example, note the yellow and black bands in wasps and the black and white stripes in skunks). The Bumblebee Poison Frog (*Dendrobates leucomelas*) is so called because of its distinctive yellow and black bands.

Some frogs combine both cryptic coloration and bright colors to warn predators. For example, when at rest on a leaf, the Red-eyed Treefrog (*Agalychnis callidryas*, see page 210) of Costa Rica closes its eyes, folds in its legs, and relies on its green coloration to match the background. If disturbed, it opens its enormous red eyes and unfolds its legs to display bright blue patches on its thighs and flanks. This flash coloration may momentarily startle a predator, giving the frog time to escape.

↗ Found from Nicaragua to northeast Colombia, the Green and Black Poison Frog (*Dendrobates auratus*) is among the largest of the poison frogs (up to 1½ in/4 cm), and is popular in the pet trade due to its size and variety of color morphs.

→ A Müllerian mimic of the Brilliant-thighed Poison Frog (*Allobates femoralis*), the Painted Ant-nest Frog (*Lithodytes lineatus*) is so named due to its association with highly aggressive Leaf-cutter Ants (*Atta cephalotes*) in whose nests it is regularly found. The frogs mimic the chemical signal of the ants which prevents them from being attacked.

CAMOUFLAGE, WARNING COLORATION, AND MIMICRY

MIMICRY

The more conspicuous the colors, the more quickly predators will learn to avoid prey with those colors. In these circumstances, it may be beneficial for some perfectly palatable prey to mimic those that are noxious and avoided by predators. This is known as Batesian mimicry. However, this will usually only work if the mimics are much less common than the "model," otherwise the predator will learn that most of the brightly colored individuals are perfectly edible. An alternative form of mimicry is when different species all conform to a common color, which warns predators that they should not be attacked. Known as Müllerian mimicry, one of the best examples is found in the appropriately named Imitating Poison Frog (*Ranitomeya imitator*, see page 218).

↖ The Yellow-banded Poison Frog (*Dendrobates leucomelas*) has striking yellow/orange and black aposematic coloration (it is sometimes called the bumblebee frog). The toxins which accumulate in its skin are derived from a diet of ants.

↑ The Ecuador Poison Frog (*Ameerega bilinguis*) has a reddish back and shows bright yellow flashes on its upper arms and thighs. Its belly is also strikingly patterned blue with black lines and blotches, advertising its toxicity. This species also has a Batesian mimic, the Sanguine Poison Frog (*Allobates zaparo*).

DEFENSES AGAINST PREDATORS AND PATHOGENS

AGALYCHNIS CALLIDRYAS

Red-eyed Treefrog

Green eggs and tenacious tadpoles

SCIENTIFIC NAME:	*Agalychnis callidryas*
FAMILY:	Hylidae
LENGTH:	2¼–3 in (55–77 mm)
LIFE HISTORY:	Terrestrial eggs (around 40 in clumps) and adults; aquatic tadpoles
NOTABLE FEATURE:	Its striking coloration makes this species one of the most recognizable frogs in the world
IUCN RED LIST:	Least Concern

Distributed from Mexico in the north to Panama and Columbia in the south, the Red-eyed Treefrog is found in tropical lowland and montane forest in the vicinity of temporary or permanent ponds on leaves overhanging water bodies. The arboreal eggs are laid in sticky clutches and face a range of threats, including predation by Cat-eyed Snakes (*Leptodeira septentrionalis*) and several species of wasp.

Red-eyed Treefrog eggs are laid in a mass of jelly on a substrate, such as leaves or branches, overhanging a water body. Six to ten days after the clutch is deposited, the tadpoles hatch and leave the egg mass by dropping into the water below. As few as four days after being laid, the tadpoles develop the capacity to escape predators by hatching early. If threatened, a tadpole can escape its egg in as little as 6.5 seconds, and some tadpoles even manage to hatch from their egg while in the jaws of a snake! The price paid by such early hatching tadpoles is the risk of then falling prey to aquatic predators, although it is better to avoid certain death in the jaws of a snake and run the risk of being eaten by something else in the water. Even more fascinating is that within the egg, tadpoles can distinguish between threatening and nonthreatening vibrations, and do not respond to the disturbance caused by rainfall or wind. Furthermore, the tadpoles can also survive out of the egg —and water—for as many as 20 hours, providing further opportunity for survival, with the possibility that rainfall or their own movements will propel them into the water.

With their green background coloration when at rest, metamorphosed and adult Red-eyed Treefrogs look very much like many other treefrog species. However, these frogs have extensive flash-coloring—an adaptation to distract predators when disturbed and buy time for escape—and combine bright red eyes with blue flanks interspersed by vertical yellow stripes, blue or orange on the underside of their limbs, and orange hands and feet.

→ Despite its bright coloration, the skin of the Red-eyed Treefrog does not contain toxic compounds like (for example) dendrobatid and mantellid frogs. However, it does contain biologically active peptides which may confer antimicrobial or antifungal resistance.

DEFENSES AGAINST PREDATORS AND PATHOGENS

DRYOPHYTES CHRYSOSCELIS

Southern Gray Treefrog

Death-inspired, shape-shifting, color-changing tadpoles

SCIENTIFIC NAME:	*Dryophytes chrysoscelis*
FAMILY:	Hylidae
LENGTH:	1¼–2 in (30–50 mm)
LIFE HISTORY:	Aquatic eggs (30–40 in groups) and tadpoles, terrestrial adults
NOTABLE FEATURE:	When breeding, male frogs will regularly call from the ground
IUCN RED LIST:	Least Concern

The Southern Gray Treefrog is widely distributed in the eastern half of the United States, extending into southern Canada. It shares much of this range with the closely related Eastern Gray Treefrog (*Dryophytes versicolor*) from which it is almost indistinguishable, although the two species differ in their vocalizations and the number of chromosomes (the structures that carry the genetic material).

Southern Gray Treefrog tadpoles that occur in ponds with dragonfly nymphs have longer tails with a deeper and more brightly colored tail fin than those in ponds without dragonflies. The colored tail fin serves to divert dragonfly attacks away from the vulnerable head and body of the tadpole and toward the more expendable tail, which can grow back if damaged. If dragonflies colonize a pond, the tadpoles can change their body shape and tail color in as little as 14 days. After dragonfly nymphs have eaten tadpoles, they release a chemical into the water to which the remaining tadpoles respond by changing their body shape and tail color. This suggests that the surviving tadpoles use the remains of others that have been excreted into the water after digestion by the dragonflies as a signal to change their form. Intriguingly, Southern Gray Treefrog tadpoles also respond in the same way to chemical cues from dragonflies that have eaten the tadpoles of the Spring Peeper Frog (*Pseudacris crucifer*), an adaptive sensory cue that can lead to improved survival in both species.

→ While Southern Gray Treefrog tadpoles use their colored tail fin to divert predator attention, adult frogs rely on camouflage, also blending into their surroundings by changing their background coloration. Although not as dramatic as the Red-eyed Treefrog, they also have flash coloration (yellow-orange patches) under their forearms and around the hindlegs, which are visible when the frog moves or jumps from a resting position.

DEFENSES AGAINST PREDATORS AND PATHOGENS

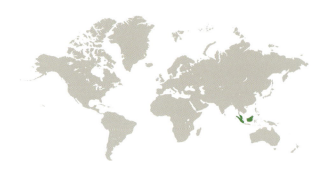

PELOBATRACHUS NASUTUS

Long-nosed Horned Frog

Forest-floor master of disguise

SCIENTIFIC NAME:	*Pelobatrachus nasutus*
FAMILY:	Megophryidae
LENGTH	4–4¾ in (100–120 mm)
LIFE HISTORY	Aquatic eggs (clusters of 500–1800) and tadpoles; terrestrial adults
NOTABLE FEATURE	Fleshy protuberances on the nose and above each eye give the appearance of horns
IUCN RED LIST	Least Concern

A member of the largest clade of archaeobatrachian frogs—the Pelobatoidea—this highly distinctive anuran is widely distributed in the forests of the Malaysian Peninsula, Borneo, and Sumatra, and is also found in Singapore. Its distinctive leaflike appearance provides the perfect camouflage to conceal it from predators among the leaf litter of the forest floor.

The Long-nosed Horned Frog (also known as the Long-nosed Horned Toad or the Malayan Horned Frog) is strikingly patterned. It has angular ridges and lines, and a host of color variations, in addition to a horned nose and upper eyelids, which are not actually horns but fleshy projections. When at rest on the forest floor, the angular horns, dorsal ridges, and even the mouth of this frog serve to disrupt the outline of the body, making them almost impossible to spot, unless they move. Their coloration even matches the browning and mottled patterning of leaf litter: the ridges along the back and between the eyes on top of the head look like the venation on a fallen leaf.

Although the horns of this frog are not constructed from bone, this species does in fact have bone in other areas of its skin. Many small, irregularly shaped bony plates, known as osteoderms, are present in the skin on the back of the Long-nosed Horned Frog. It is thought that these provide a degree of protection from predators by making the skin more robust.

Tadpoles of the Long-nosed Horned Frog are also distinctive in appearance, and have mouthparts shaped like a giant pair of upturned lips that form a funnel. These are highly adapted for filter-feeding at the surface of the water, in leaf drift in the more sedate sections of flowing streams. On average, tadpoles take approximately three months to reach metamorphosis, but froglets do not develop their characteristic horns until several weeks after leaving the water.

→ As well as skin folds that give a ridged leaflike effect, and horns above the eyes and nose, the Long-nosed Horned Frog also has scattered fleshy projections—tubercles—over the upper surface of its body, including the legs. These further help to break up the frog's outline in the leaf-litter, enhancing its camouflage.

PRISTIMANTIS MUTABILIS

Mutable Rainfrog

The shape-shifting punk rocker frog

SCIENTIFIC NAME:	*Pristimantis mutabilis*
FAMILY:	Strabomantidae
LENGTH:	½–1 in (15–23 mm)
LIFE HISTORY:	Unknown, but other *Pristimantis* frogs are terrestrial breeding direct developers with parental care
NOTABLE FEATURE:	Spiky, tubercular skin that can become smooth in a few minutes
IUCN RED LIST:	Endangered

At nearly 600 species and with new species being described all the time, *Pristimantis* from Central and South America has more species than any other vertebrate genus. The Mutable Rainfrog was discovered in 2015 and is known only from a few sites in the western Andes of Ecuador. Although many frogs can change color according to environmental conditions, this was the first frog discovered that can change the texture of its skin over a short period of time.

The Mutable Rainfrog is nocturnal, spending much of its time hiding in vegetation or moss. It is strikingly marked with contrasting shades of green and brown and has skin which contains numerous tubercles on the upper body and legs. These tubercles are particularly pointy above the eyes, and their appearance has given the Mutable Rainfrog its alternative name of "punk rocker frog." When the frog was discovered, some individuals were placed in plastic boxes, and within a few minutes had lost their spiky appearance—the skin had become almost completely smooth. On being returned to a damp, mossy environment the tubercles quickly reappeared. Although the reasons for this change in skin texture remain unknown, it may be linked to maintaining camouflage in different environments, with a skin that develops tubercles more closely matching a damp, mossy habitat. The plasticity that the Mutable Rainfrog shows in changing from tubercular to a smooth skin demonstrates just one of the challenges with using physical characteristics to classify species.

Changeable skin
The Mutable Rainfrog is capable of changing the texture of its skin from tuberculate to smooth in little more than five minutes. Exactly how it is able to do this is at present unknown, although it is thought to involve the transport of fluid to and from structures in the skin.

Tuberculate skin

Smooth skin

→ Differential skin texture in frogs is usually only observed between males and females, but in the Mutable Rainfrog, and one other related species (*Pristimantis sobetes*), this feature is not only present in both sexes, but it is a trait that these frogs can seemingly use as and when required, most likely for camouflage.

DEFENSES AGAINST PREDATORS AND PATHOGENS

RANITOMEYA IMITATOR

Imitating Poison Frog

Appearances can be deceptive

SCIENTIFIC NAME:	*Ranitomeya imitator*
FAMILY:	Dendrobatidae
LENGTH:	½–1 in (15–22 mm)
LIFE HISTORY:	Aquatic eggs (perhaps five in a clutch) and tadpoles; terrestrial adults; parental care
NOTABLE FEATURE:	The only known monogamous amphibian
IUCN RED LIST:	Least Concern

The Imitating Poison Frog is so called because it imitates the appearance of three other related species: the Crowned Poison Frog (*Ranitomeya fantastica*), the Variable Poison Frog (*R. variabilis*), and Summers' Poison Frog (*R. summersi*). These species differ in appearance, where areas of lighter coloration form either spots, bands, or stripes. Co-occurring populations of the Imitating Poison Frog therefore replicate the color patterns of whichever of the three species they happen to be living alongside. However, the call of the Imitating Poison Frog remains quite different to the species it is imitating.

All of the frogs within the genus *Ranitomeya* have toxic skins, and their bright coloration—distinctive yellow, orange, or red body markings on a black background, and some with blue feet—advertises this fact to predators. The Imitating Poison Frog is therefore able to mimic the coloration of other species with which it coexists. This means that the different species gain collective protection from predators by having the same aposematic coloration. It is a phenomenon known as Müllerian mimicry, which is common in some yellow- and-black banded insects but rarer in frogs. If a predator eats one species and finds it distasteful, it will subsequently also avoid eating species with a similar pattern.

Among frogs, it is rare that both the male and female have parental care of eggs, tadpoles, and froglets—the task is usually done by one or other of the sexes. With Imitating Poison Frogs, however, both parents care for their offspring. Fertilized eggs are deposited in phytotelma, where they undergo initial development. Around two weeks later, the male frog transports the pair's tadpoles to a different plant, which likely confers a benefit (predator avoidance, or minimizing disease, for example). However, the best tadpole nurseries for these frogs are small, but nutritionally poor. To ensure their tadpoles have enough good-quality food, the male frog makes an egg-feeding call to the female, which encourages her to provision their tadpoles with an unfertilized egg. This coordinated strategy of parental care is thought to be driven by the nutrient-poor nurseries in which the tadpoles are raised. This has resulted not only in both parents of this species caring for the eggs and tadpoles, but also the evolution of pair bonding and monogamy as crucial factors in their reproductive success.

→ The spotted morph of the Imitating Poison Frog is a Müllerian mimic of the Variable Poison Frog (*Ranitomeya variabilis*). Both share a toxic defense and benefit from highly similar aposematic coloration.

THE UPS & DOWNS OF FROG POPULATIONS

THE UPS AND DOWNS OF FROG POPULATIONS

Survival strategies

The number of frogs in a population depends on their life history and how this relates to climate, predation, competition, disease, and features of the wider landscape. Interactions between these different drivers mean that many frog populations undergo regular but natural fluctuations in population size. While most of what we know about survival in frog populations has been derived from observations of explosive-breeding temperate species, similar patterns are likely in tropical frogs with different life histories.

Ultimately, how many adult frogs breed each year depends on how many eggs and tadpoles have survived in previous years. In turn, the survival of eggs and tadpoles is driven by competition for space and food, and the impacts of predators and disease. High densities increase competition between tadpoles and can draw in predators attracted to aggregations of prey. Likewise, disease-causing pathogens are more easily transmitted in a crowd. On the other hand, if a pond desiccates, then all eggs and tadpoles will die, irrespective of whether they are at high or low density.

→ A blue male Moor Frog (*Rana arvalis*) in amplexus with a female. As well as being vulnerable to predators at this time, breeding is energetically demanding, and frog populations can suffer high mortality.

SURVIVAL STRATEGIES

The reason many frogs lay thousands of eggs at a time is an evolutionary strategy to ensure that at least a few offspring survive to adulthood, as most tadpoles will fail to reach metamorphosis. On average, a pair of frogs that produces, say, 10,000 eggs in their lifetime, only needs two of those eggs to survive to adulthood in order to contribute to maintaining a stable population. In fact, in some years no eggs or tadpoles may survive if the pond dries up. Equally, if there is a drought, the frogs may skip breeding altogether. Consequently, if a frog lives for five years—taking two years to reach adulthood after which there are three years of breeding opportunities—it may only need one of those years to produce enough offspring to eventually join the adult population and ensure its long-term stability.

Total failure or loss of eggs or young, or skipping a breeding year altogether, may therefore not be as catastrophic as it sounds if the frog survives long enough to have multiple breeding opportunities to offset the losses. Understanding how long frogs live can provide clues as to how often they breed and how their populations will persist over time. Even factors such as elevation can influence longevity, as in the case of the Boreal Chorus Frog (*Pseudacris maculata*, see page 238).

"BOOM AND BUST" REPRODUCTION

In terms of long-term population trajectories, this rather unpredictable pattern of reproductive success leads to a highly fluctuating "boom and bust" dynamic. In "boom" years a lot of tadpoles make it to metamorphosis, leading to population increases when they become adult; in "bust" years the population plummets because of zero recruitment to the adult population. Such populations may not show any sort of constancy in size, but so long as the "boom" years offset the "bust" years, in the long-term the population may be stable.

Research on the breeding of the Ornate Chorus Frog (*Pseudacris ornata*) at a pond in South Carolina, from 1979 to 1990, revealed wide year-to-year fluctuations in population sizes, and despite a 30-fold decrease in female frogs from 1983–1989, over the

HOW OLD IS THAT FROG?

Frogs that see the winter out in hibernation undergo slowed bodily functions and reduced growth. Such slowed growth results in growth rings, or Lines of Arrested Growth (LAGs), being laid down in growing bones. Therefore, in the same way that counting the rings in a cut tree stump will provide the age of the tree, so counting the growth rings in frog bones can indicate how many winters that individual has passed through—this is known as "skeletochronology." Although the rings are clearest in the long bones of the arm (humerus) and leg (femur), they are also laid down in the fingers and toes. As a small part of a digit can be removed without killing the frog, this method can be applied to large numbers of frogs in a population. The information provided can help determine if the frogs are part of a young, expanding population or an aging, declining one.

Lines of arrested growth

SURVIVAL STRATEGIES

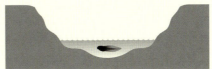
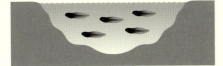

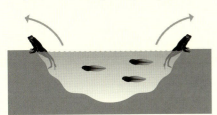
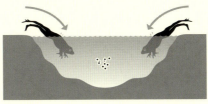

Comparison of reproductive success in wet versus dry years

Survival of tadpoles to metamorphosis and their recruitment to the wider population is supported in wet years. In dry years, frogs may breed but tadpoles do not survive to metamorphosis. Providing adult frogs live long enough to breed for another year, they may return when favorable conditions refill the pond.

↓ The Marbled Snout-burrower (*Hemisus marmoratus*) undergoes "boom and bust" population fluctuations, with the survival of adults and tadpoles highly dependent on patterns of rainfall.

whole 12-year monitoring period there was no overall trend. These wide fluctuations could be accounted for by periodic droughts. In another North American study spanning some 20 years, Fowler's Toads (*Anaxyrus fowleri*) tended to be older during population dips than during population increases. This is because the increases reflect greater numbers of younger toads joining the adult population during those periods. As this doesn't happen during the population dips, the toads have a higher average age. However, adult survival is also important, and biotic factors such as a lack of rainfall can negatively impact even ecologically adapted frog populations (for example, the Marbled Snout-burrower, *Hemisus marmoratus,* see page 240).

225

↑ Found on the island of Borneo, the Lowland Litter Frog (*Leptobrachium abbotti*) is a member of the Megophryidae and a forest floor leaf-litter specialist. The orange/red coloration of the ridge of skin, which extends from the nose to behind the head and jaw, fades to become less vivid as these frogs move into adulthood.

SURVIVAL STRATEGIES

↑ Endemic to Papua New Guinea, the function of the rostral appendage in males of the Northern Pinocchio Treefrog (*Litoria pinocchio*) is currently unknown, but this rostral spike is known to inflate upward when calling, and therefore may be important for reproductive competition and mate selection.

THE UPS AND DOWNS OF FROG POPULATIONS

The importance of dispersal

Although many frogs may have several opportunities to lay clutches of eggs during their lifetime, unpredictable events — such as prolonged droughts or fires — may mean that most frogs in the population will fail to produce a single froglet before they die. Are such populations doomed to extinction?

Frog populations occupying ponds that repeatedly undergo droughts or other disasters can be rescued by the arrival of frogs from another population, thereby sustaining their numbers. This requires frog populations to be within dispersal range and for suitable habitat to exist between them to allow for individual dispersal (see Colonizing New Breeding Sites, page 144). If one population goes into decline, it may be compensated for by dispersal from a neighboring population which is increasing and producing frogs that move to new breeding sites.

← The Mallorcan Midwife Toad (*Alytes muletensis*) breeds in plunge pools along rocky gorges. Following rains, the torrents flow and sweep tadpoles downstream.

MOLECULAR GENETICS AND FROG POPULATIONS

Obtaining a sliver of tissue from a tadpole's tail, a spot of blood from a vein, or a few skin cells from a frog's mouth can provide a wealth of information about that individual. When many frogs from the same population are sampled and the data compared to those from other populations, it is possible to reconstruct both their recent and longer-term histories. By exploring genetic diversity, we can determine whether a population may have previously declined to critically low numbers (a population bottleneck), or if it had always been small, such as when a limited number of individuals form a new population, but no subsequent immigration takes place (known as the founder effect). Both can negatively impact long-term population viability through inbreeding, and this too can be detected by assessing genetic diversity, as can landscape-level barriers to dispersal and whether populations have a metapopulation structure.

If a population is sampled over time, ideally accounting for several generations, the degree of genetic change observed can be used to estimate the effective population size. Simply defined, this is a measure of the number of frogs contributing genetic material to the population. As not all frogs will breed each year, the effective population size is invariably much smaller than the census population size obtained by traditional counts or observations. For example, the effective population size in Dusky Gopher Frogs (*Lithobates sevosus*, see page 242) has been estimated as 50 percent of the census population size, given the number of adult frogs captured.

Using DNA and eDNA to monitor frogs and frog communities

Genetic material from frogs can be derived from biological samples such as tissue (DNA) or collected as cells that are shed into the environment as frogs move within it, and then contained in samples of soil or water (eDNA). The resultant material can be analyzed to determine how much DNA of a given species or community might be present in the environment (quantification), or it could be used to verify the taxonomic identity of species present (metabarcoding) and understand community composition.

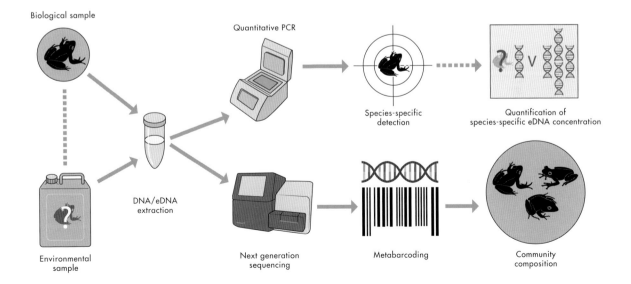

THE METAPOPULATION CONCEPT

Whether a cluster of ponds in a group of fields, a group of bromeliads on a mountain top, or a series of water-filled hollows in trees in a forest, many frogs have natural populations scattered across a landscape. Such a "population of populations" is referred to as a metapopulation. The key element of a metapopulation is that the component populations occupy habitats that are linked by dispersal. If frogs disperse from a population, they are emigrating, while frogs moving into a different population are immigrating. A population producing an excess of frogs that disperse to other areas is a source, while one where recruitment fails is a sink. However, such is the variation of climate and other population drivers across a habitat landscape that over time a source may become a sink, and vice versa. Suitable habitat corridors that provide appropriate cover, such as long grass, waterways, or patches of connected forest, are needed to link the populations and facilitate dispersal.

If reproduction fails for several years or a disaster strikes one or more of the connected populations, immigrants from a neighboring population may move in and replenish frog numbers. A metapopulation of pond-breeding frogs therefore requires three components: good-quality breeding sites; good-quality terrestrial habitat to support the terrestrial phase; and good-quality connecting habitat between the ponds. All three are needed to ensure a sustainable, long-term population of frogs.

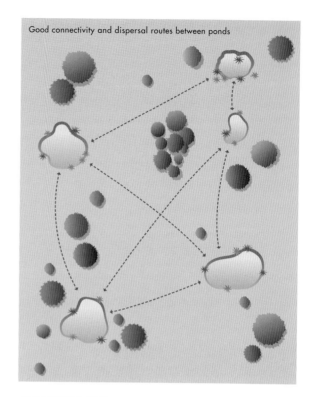

Good connectivity and dispersal routes between ponds

No connectivity and dispersal routes between ponds

Scattered populations
A viable population of frogs usually requires several ponds with surrounding terrestrial habitat, and corridors to allow dispersal (top). Even if ponds and terrestrial habitat are retained within a development, barriers to dispersal may mean that the population goes into decline (bottom).

BARRIERS TO OVERCOME

If the long-term viability of a frog population requires movement of individuals between sites to compensate for natural ups and downs, what barriers to dispersal must be surmounted? Indeed, how can dispersal within a complex landscape be identified? These are fundamental questions that frog biologists must tackle to ensure frogs can survive within the wider landscape.

The most obvious way to monitor movements is through tracking individuals. However, there are practical challenges in achieving this (see Chapter 5), and a few frogs with radio-transmitters may not reflect dynamics of the wider population where dispersal may be biased toward males or females, or newly metamorphosed juveniles that are problematical to track. Fortunately, molecular genetic methods are informative here. Individual frogs, and often frog populations, have a unique genetic signature. The degree to which genetics are shared between individuals and populations provides information about how related frogs are within and between different populations. Therefore, genetic information can give insights into the historical movement of frogs within the landscape.

Mountains often provide barriers to dispersal between populations. For example, genetic analyses of Columbia Spotted Frogs (*Rana luteiventris*), in Idaho and Montana, and Mallorcan Midwife Toads (*Alytes muletensis*, see page 102) both showed that populations were isolated by mountain ridges. Both species have individuals that will use several pools within their respective habitats, and in the case of the Mallorcan Midwife Toad most dispersal is downstream. This probably occurs after rains when the torrents flow between plunge pools and wash the tadpoles out. The fragmentation of natural habitats by human activity creates barriers to dispersal with consequences for the viability of frog populations.

↖↑ Although Mallorcan Midwife Toads (left) are found in several gorges across the Serra de Tramuntana mountain range (right), there is relatively little dispersal over the inhospitable terrain between these narrow valleys.

Implications for conservation of populations

Natural fluctuations in population size present challenges for tackling conservation problems. The difficulty in identifying a population trend means that frog populations require continual study over multiple years—at least several generations—to distinguish a genuine decline from a short-term blip.

Although it is very concerning when amphibians undergo catastrophic declines and die-offs due to, for example, disease (see Chapter 9), most extinctions are protracted and largely imperceptible. Repeated years without successful reproduction or immigration will mean that the population becomes increasingly comprised of older animals that will eventually die. Such populations are doomed to a slow extinction and have therefore been termed "the walking dead."

Although frogs have several built-in strategies to help them deal with the vagaries of climate and occasional disasters—large clutch sizes, repeated multi-year breeding, prolonged breeding periods, dispersal—in the face of climate change these strategies are delicately poised.

↗ Common Toads (*Bufo bufo*) can maintain their population by reproducing in large numbers across a few locations in a given area.

← Understanding frog population dynamics is challenging because many species, especially direct developers such as the Fiji Treefrog (*Cornufer vitiensis*)—seen here guarding its young—do not converge on breeding sites en masse, making them even more difficult to monitor.

→ Populations of Natterjack Toads (*Epidalea calamita*), which had declined in number to the extent that they lost significant levels of genetic diversity, have since been restored thanks to the introduction of new genetic material.

IMPLICATIONS FOR CONSERVATION OF POPULATIONS

If droughts become a longer and more intense feature of the landscape, and frogs from source populations are unable to disperse due to anthropogenic or other barriers, then populations will decline. Distinguishing genuine population declines from natural fluctuations can therefore be difficult and multiple factors need consideration (see the Mountain Yellow-legged Frog, *Rana muscosa*, page 244, for example).

DIFFERING RESPONSES IN SPECIES

Species respond in different ways to environmental changes even if they occupy the same areas. For example, in Great Britain the European Common Frog (*Rana temporaria*, see page 32) and European Common Toad (*Bufo bufo*, see page 104) have overlapping ranges. Although they sometimes breed in the same ponds, Common Frogs prefer small ephemeral ponds while Common Toads use larger, rural ponds, and due to their tadpoles' toxicity coexist with fish rather better. Common Frogs seem to maintain higher levels of genetic diversity because a population can be distributed between a large number of small ponds. In contrast, although Common Toad populations are often much larger than Common Frog populations, they are often focused on single breeding areas that are more widely dispersed within the landscape.

Due to the greater distances between Common Toad breeding sites compared to those of Common Frogs, toad populations are more easily isolated by road building and other developments that fragment the habitat. The differences in habitat preferences between the two species may therefore be related to differences in genetic diversity, and, ultimately, provide clues as to why Common Toads are mysteriously declining in areas where Common Frogs continue to do well. In the Natterjack Toad (*Epidalea calamita*, see page 246) there is evidence that an introduction of novel genetic material—genetic rescue—can help restore a declining population.

THE UPS AND DOWNS OF FROG POPULATIONS

How many frogs in a pond?

Understanding how many populations of frogs there are, and how many individual frogs in a population, is fundamental to understanding population trends and determining ways of ensuring their future. If the goal is simply to conclude whether frogs are present or not at a pond or in a patch of forest, then direct visual surveys or listening for calls of the target species may help establish this. However, an indication of the number of frogs present may provide more useful information.

Frogs are usually easier to count if the species aggregates at breeding sites, and fortunately they are often so focused on the mating game that they are oblivious to those doing the counting. A drift fence and pitfall traps erected around a breeding site will intercept frogs migrating to-and-fro and can assist with establishing numbers. Frogs run over by cars on adjacent roads can (sadly) also provide an indication of the numbers on the move. However, many individuals will be missed irrespective of the counting method, so the numbers obtained may not be good indices of population size.

To obtain accurate estimates of population size requires rather more sophisticated measures. This often means identifying individual frogs through their color patterns, or by marking them with a tag. By carrying out repeated marking sessions and observing the numbers of marked frogs recaptured in each session, it is possible to estimate the population size. Indeed, if carried out over a long period of time, such capture-mark-recapture exercises can also reveal survival rates, and gains and losses to each population.

↑ The number of metamorphosing frogs and toads fluctuates from year to year. The number that return to breed will depend on how many survive to adulthood, and how many disperse between the breeding sites.

← Although often easy to count at a breeding site, the numbers of frogs observed at a single point in time may only represent a fraction of those present in the wider population.

USING TECHNOLOGY

Sometimes all that is needed to count frogs is a strong torch, a raincoat, and a willingness to go out at times of the day or night when they are likely to be active. But longer-term studies may require more sophisticated equipment. Camera traps can be deployed at breeding sites and left in place for several days or weeks to take photographs of any frogs that show themselves. Such cameras have even been used to monitor tiny Itambé Bromeliad Frogs (*Crossodactylodes itambe*) that spend their entire lives in bromeliads at the top of a mountain in Brazil.

Similarly, bioacoustic recorders can be deployed at a breeding site or across a given habitat and left to passively record frog calls—particularly effective for difficult-to-observe, behaviorally cryptic frogs that call from leaf litter on the forest floor, or from many meters up in the canopy. New computational approaches incorporating machine learning can be applied to classify animals from hundreds of thousands of camera trap images or extract the vocal signature of target species from massive bioacoustics datasets.

If it is important to identify individual frogs in a population, computational solutions can also process and match up photographs of known animals. This method may be appropriate for frog species with individually distinctive color patterns that can be conveniently captured on camera. For those more challenging to identify by their markings, a variety of tags are now available. Passive Integrated Transponders

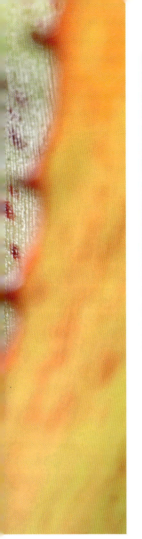

↖ Many frogs use bromeliads, but few are truly bromeligenous—spending almost their entire lives within these plants. Yellow Heart-Tongued Frogs (*Phyllodytes luteolus*) of eastern coastal regions of Brazil depend on bromeliads for their entire life cycle, and male frogs will fight each other for access to these critical breeding resources.

↗ Tiny microchips implanted under the skin of this Oregon Spotted Frog (*Rana pretiosa*) allow the identification of individuals within the population.

(or PIT tags) about the size of a grain of rice are easily injected just beneath the skin. Each tag carries a unique barcode which can be read each time the frog is captured using a supermarket-style barcode reader. Such tags have also proved useful for monitoring predation, as they can be picked up in the stomachs of predators that have eaten frogs. One captured and scanned Grass Snake (*Natrix helevtica*) had three PIT-tagged European Common Frogs in its stomach!

Visible Implanted Elastomers (VIE tags) are colored tags that can be implanted into very small frogs, and even tadpoles. If tadpoles are marked, VIE tags have the advantage of being carried through metamorphosis, so froglets can be identified. In the Channel Island of Jersey, these tags were used to identify metamorphs of Agile Frogs (*Rana dalmatina*) raised in captivity and then released into a pond to complete development.

PSEUDACRIS MACULATA

Boreal Chorus Frog

A wide-ranging North American species

SCIENTIFIC NAME:	*Pseudacris maculata*
FAMILY:	Hylidae
LENGTH:	1–1½ in (25–35 mm)
LIFE HISTORY:	Aquatic eggs (500–1,500 in small clumps) and tadpoles; terrestrial adults
NOTABLE FEATURE:	Shorter legs make this frog a hopper not a leaper
IUCN RED LIST:	Least Concern

The Boreal Chorus Frog has a varied background coloration that ranges from greens to browns and grays. The back usually shows three dark dorsal stripes that may be broken and/or spotted. Found across central and eastern North America, this species has perhaps the widest distribution among the "striped chorus frogs" to which it belongs. Reliable identification in these taxa may therefore require combined analysis of calls, morphology, genetics, and geographical information. The extensive range of the Boreal Chorus Frog means that there can be considerable variation in population biology and life history.

Historical datasets can provide useful information about frog populations, but comparisons with more recent data are often difficult due to differences in data collection methods. Reanalysis of a nine-year study of Boreal Chorus Frogs in an upland pond in Colorado, starting in 1963, has recently revealed important information about the frogs at that time. Using statistical models, the researchers found that the frogs lived for five to seven years, with 67–79 percent of the frogs surviving their first year. This is probably a longer lifespan than occurs at lower elevations. As elevation increases, Boreal Chorus Frogs hibernate for longer, perhaps for up to six months of the year. This slowing down of bodily functions for long periods may facilitate a longer life, and suggests that adult survival is more important for population stability and growth than recruitment of frogs through regular breeding, at least at higher, cooler elevations.

Predation is a primary influence on frog populations, and the level of breeding success can be driven by its relative impacts. Many frog species, including the Boreal Chorus Frog, select breeding ponds based on their absence of fish, for example. Tadpoles also avoid predation by adopting behaviors such as schooling, but learning, and in particular social learning, is relatively unknown in amphibians (although this is perhaps due to a lack of research). The Boreal Chorus Frog and Wood Frog (*Lithobates sylvaticus*) are sympatric across a significant proportion of their respective range, as are Tiger Salamanders (*Ambystoma tigrinum*), which feed on tadpoles of both species. When experimentally exposed to the odor of predatory Tiger Salamanders, naïve Boreal Chorus Frog tadpoles appear to show no reaction. However, when paired with predator-experienced Wood Frog tadpoles, they learn to recognize salamander odor and display a fright response. This kind of social learning may therefore be crucial for increasing individual survival within frog populations and amphibian communities.

→ The Boreal Chorus Frog can be found in a range of habitats, including grasslands, marshes, farmland, and urban areas.

HEMISUS MARMORATUS

Marbled Snout-burrower

A determined parent

SCIENTIFIC NAME:	*Hemisus marmoratus*
FAMILY:	Hemisotidae
LENGTH:	1¼–1¾ in (30–45 mm)
LIFE HISTORY:	Terrestrial eggs (88–242 in a subterranean nest) and adults; tadpoles aquatic at later stages of development; parental care
NOTABLE FEATURE:	One of the few frogs that burrows head-first
IUCN RED LIST:	Endangered

The breeding of shovel-nosed frogs such as *Hemisus marmoratus* is integrally tied to rainfall, with fewer adult frogs surviving in dry years compared to wet years. Tadpole survival can be highly variable, and the most important driver of population growth for the Marbled Snout-burrower is survival to metamorphosis and juvenile recruitment.

In some years, freshwater turtles colonize Marbled Snout-burrower breeding pools and consume large numbers of tadpoles. Predation combined with the impact of dryer years leads to a drastically fluctuating "boom and bust" population cycle. After a "bust" year, it can take three years for the numbers of frogs to recover to previous levels. However, one way that the Marbled Snout-burrower manages the many factors that can influence tadpole survival, and therefore population recruitment, is through investment in their offspring. This starts at the moment breeding activity begins—when the rains arrive and eggs are laid in a subterranean nest chamber excavated by the female. Having been laid, the eggs are surrounded by a layer of empty egg capsules and covered with a protective film. These provisions likely protect the eggs from moisture loss, and the female will also actively defend her offspring from predators.

Within the burrow, the eggs hatch, and the tadpoles may remain in situ for several days, or for up to two months. When conditions allow, two or perhaps three modes of tadpole dispersal from the nest chamber may take place. In one example, the burrow may be dug in an area that becomes inundated soon after the initial rains, meaning that the nest chamber floods, and tadpoles are effectively washed into the new water body. If nest chambers are close to newly filled ponds, female frogs can scrape out a superficial mudslide, providing a direct route for the young to follow and enter the water. However, in a remarkable mode of tadpole transportation from nest chambers to more distant water bodies, the female frog will extend her hindlimbs and glue them together to maximize her body surface area. Then, with more than 100 tadpoles attached all over her body and now immobile legs, she pulls herself along using just her forelimbs until reaching water, whereupon the tadpoles either remove themselves, or they are dislodged by the female's vigorous kicking and swimming. Combined, these multiple modes of condition-dependent parental care mean that the Marbled Snout-burrower can maximize the survival chances of its offspring.

→ The Marbled Snout-burrower is an almost entirely subterranean frog with a diet of ants and termites (termed myrmecophagy). Given their infrequent appearance on the surface, it is likely that they forage underground, although this behavior is yet to be confirmed.

THE UPS AND DOWNS OF FROG POPULATIONS

LITHOBATES SEVOSUS

Dusky Gopher Frog
A southern US endemic

SCIENTIFIC NAME:	*Lithobates sevosus*
FAMILY:	Ranidae
LENGTH:	2–3¾ in (50–95 mm)
LIFE HISTORY:	Aquatic eggs (500–7000 in clumps) and tadpoles. Terrestrial adults
NOTABLE FEATURE:	So called as it frequently uses gopher tortoise burrows as refugia
IUCN RED LIST:	Critically Endangered

Now confined to a single wild population in Mississippi, this darkly colored and rather warty frog may be the rarest anuran in North America. Historically, the Dusky Gopher Frog was widespread across the southern coastal plain of the United States, but it has declined due to habitat change and fragmentation, predation, drought, pollution, and disease.

In 2001, the number of Dusky Gopher Frogs had dwindled to an estimated 100 individuals that were known to breed in just two ponds in Mississippi. As a longleaf pine forest specialist, the species' precipitous decline was largely driven by habitat loss and alteration, which has today reduced this forest type to less than two percent of its former pre-settlement cover. Due to the frogs' elevated extinction risk, eggs were collected and tadpoles raised to metamorphosis in outdoor tanks (a process known as head-starting), with the aim of establishing a captive population, and providing stock to aid the recovery of frogs in the wild.

Despite further efforts at several zoos the Dusky Gopher Frog did not readily breed in captivity, and in 2008 an attempt was made by Memphis Zoo to produce tadpoles using assisted reproductive technologies (ARTs, see Chapter 9), namely in vitro fertilization (IVF). The frog protocol they designed proved successful, and over the following five years the growth rate of the population increased by 91 percent.

Dusky Gopher Frogs were now available for reintroduction, but their ecology and behavior, and most importantly, survivorship, was not well understood. It was therefore crucial to monitor released individuals to gain these valuable insights. For this, 53 juvenile frogs were fitted with a radio-tracker and released. Fortunately, survivorship was high (76 percent), and frogs dispersed naturally, illustrating that ARTs can be used to increase and manage populations of threatened frogs.

In 2020, Dusky Gopher Frogs bred naturally in captivity for the first time, producing more than 11,000 eggs and 7,887 hatchling tadpoles. This recent success was largely due to an improved replication in captivity of the frog's natural habitat. Although the population now appears to be on the increase (514 mature individuals as of 2021), the species undoubtedly lost genetic diversity having declined to around 100 individuals at the turn of the century. Careful management using selective breeding will therefore continue to play a critical role in maintaining this species' remaining genetic diversity.

→ Previously thought to be a subspecies of the Carolina Gopher Frog (*Lithobates capito*), the Dusky Gopher Frog was elevated to species status in 2001.

RANA MUSCOSA

Mountain Yellow-legged Frog

The mossy colored frog

SCIENTIFIC NAME:	*Rana muscosa*
FAMILY:	Ranidae
LENGTH:	1½–3½ in (40–90 mm)
LIFE HISTORY:	Aquatic eggs (100–300 in clumps) and tadpoles, semiaquatic adults
NOTABLE FEATURE:	Underside of hind legs are various shades of yellow
IUCN RED LIST:	Endangered

Part of a species complex that also includes the closely related Sierra Nevada Yellow-legged Frog (*Rana sierrae*), the Mountain Yellow-legged Frog—also known as the Sierra Madre Yellow-legged Frog—is endemic to a few mountain ranges in California. Formerly abundant in protected areas, it has declined by over 90 percent since the 1960s due to a range of threats, including habitat loss, fires, drought, floods, introduced fish, and disease. It is now confined to less than ten isolated populations.

Analyses of eight of the remaining populations of frogs showed wide variations in population size over two decades. This is down to a high level of unpredictability in disturbances, which can cause population crashes in different places at different times. However, the overall trend was of regional decline. This means the populations have a high risk of extinction in the coming decades. So, with few frogs in few populations, should conservation efforts focus on boosting numbers in the existing populations or establishing brand-new populations through the release of captive bred frogs? Modeling of the remaining populations suggests that in addition to arresting drivers of declines, any reintroductions of frogs should focus on establishing new populations rather than bolstering existing ones. Nevertheless, finding suitable areas in which to establish such populations remains a challenge.

→ The Mountain Yellow-legged Frog is rarely observed far from the water's edge, and can be found in a variety of water bodies, from streams and rivers to lakes and pools at elevations of 1000–13,000 ft (300–4,000 m).

THE UPS AND DOWNS OF FROG POPULATIONS

EPIDALEA CALAMITA

Natterjack Toad
The running toad

SCIENTIFIC NAME:	*Epidalea calamita*
FAMILY:	Bufonidae
LENGTH:	1¾–3¼ in (45–80 mm)
LIFE HISTORY:	Aquatic eggs (string of up to 4000) and tadpoles; terrestrial adults
NOTABLE FEATURE:	Sandy soil specialist
IUCN RED LIST:	Least Concern

Widespread in central and southwestern Europe, in Great Britain the Natterjack Toad is on the edge of its geographical range, and is a rare species confined to warm sandy habitats, such as coastal dune systems, saltmarshes, and lowland heaths. Consequently, its remaining populations are isolated and widely scattered across the country.

A typical bufonid in appearance, the Natterjack Toad is gray to olive green in color, sometimes with darker patches and reddish dots. It can be easily distinguished from other sympatric toad species, such as the European Common Toad (*Bufo bufo*) and the Green Toad (*Bufotes viridis*) by its much shorter legs, scuttling movement (likened to a mouse), and perhaps most distinctive of all, the yellow line that runs down its back.

In Great Britain, reintroduction to restored habitats has been an important component of the conservation program for Natterjack Toads since the 1970s. Ordinarily, these have comprised translocations of both spawn and tadpoles from existing populations, although toads have also been raised in captivity for release at certain locations. Through the establishment of sustainable populations, in particular on sand dune sites, follow-up monitoring has indicated that such interventions have been highly successful.

In the 1990s, one Natterjack Toad population on the east coast of England was showing low levels of genetic diversity and poor growth and development in its tadpoles. Habitat improvements were made, but to little effect, and the root issue was subsequently identified as restricted gene flow and a loss of genetic diversity, brought about by population fragmentation and a rapid decline in effective population size (the number of individuals that are breeding in a given population). From 2003, east of England Natterjacks were supplemented with tadpoles from other populations. This boosted spawn production and breeding success, and ultimately increased levels of genetic diversity. Natterjack Toads in the east of England therefore appear to have responded positively to the introduction of novel genetic material, thereby enabling continued persistence of this population.

→ Natterjack Toads breed in shallow pools that often desiccate before their tadpoles have metamorphosed. Fortunately, Natterjack lifespan is long enough to ensure that individuals within a population may only need one "good" year to contribute to its stability.

FROGS IN THE FUTURE

FROGS IN THE FUTURE

Disappearing frogs

We are currently in an amphibian decline crisis, and there are many reasons why frogs, salamanders, and caecilians are disappearing. The drivers of declining frog populations interact with each other in complex ways, and there is huge variability in how frog species, populations, and communities are responding to threats. Now, and in the future, there are likely to be some winners as well as some losers.

Regionally, the greatest numbers of threatened amphibians occur in the Caribbean, Central America, the Tropical Andes, uplands and forests of West Africa, Madagascar, western India, and Sri Lanka. In terms of life history, terrestrial direct-developing frogs appear to be more at risk than those whose eggs hatch into aquatic tadpoles. Many frog species are highly sensitive to novel pathogens and predators, while others appear unaffected. Several species such as the Cane Toad (*Rhinella marina*), American Bullfrog (*Lithobates catesbeianus*, see page 96), and Asian Common Toad (*Duttaphrynus melanostictus*, see page 268) are so adaptable and resilient that in some parts of the world where they have been introduced they now present a serious threat to native biodiversity.

→ Known from only four populations in Ecuador, the Rio Pescado Stub-foot Toad (*Atelopus balios*) was thought to have gone extinct due to disease and habitat loss, until it was rediscovered in 2010.

FROGS IN THE FUTURE

Although the overall trend is one of decline, the drivers, their impacts, and the rates of decline differ regionally and between frog species.

Amphibians are faced with a range of threats to their survival. Those detailed here such as habitat loss, disease, and overexploitation are direct, but all can be impacted and exacerbated by other indirect drivers like human development and population growth. Likewise, threats such as climate change are difficult to mitigate and will interact with other threats. Irrespective of the cause, most frog species are today facing an uncertain future.

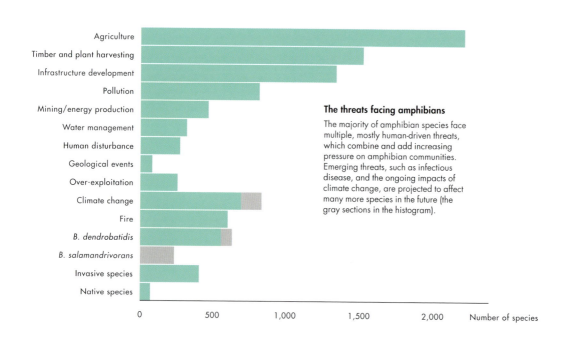

The threats facing amphibians

The majority of amphibian species face multiple, mostly human-driven threats, which combine and add increasing pressure on amphibian communities. Emerging threats, such as infectious disease, and the ongoing impacts of climate change, are projected to affect many more species in the future (the gray sections in the histogram).

252

DISAPPEARING FROGS

HISTORY OF AMPHIBIAN DECLINES

In 1989 over 1,000 researchers from all over the world gathered in Canterbury, England, for the First World Congress of Herpetology. Here, a common theme began to emerge on the status of the world's amphibians—they were showing alarming declines, and we were not at all sure why.

Some of the reports of declines predated the World Congress by several decades. In Britain, there was initial concern about the status of native frogs and newts in the 1970s, while in North America reports of declining American Toads (*Anaxyrus americanus*, see page 34) and Leopard Frogs (*Lithobates pipiens*) can be traced back to the 1950s and '60s. A puzzling factor of some declines was that they were happening in pristine areas with no obvious human pressures. Recent assessments indicate that nearly 41 percent of the world's amphibians are currently at risk of extinction, with more than 200 species already having been lost over the previous 150 years. The amphibians—around 90 percent of which are frogs—are therefore in deeper trouble than any other group of vertebrates.

← Illegal gold mining in the Amazon, and across the world, is a major threat to frog populations.

↓ Although high levels of pollution can cause direct mortality of frogs, low levels of environmental contaminants can shift interactions between frogs and their competitors and predators, affect tadpole development, and make them more vulnerable to disease.

EVOLVING THREATS:
THE CHALLENGES FACING ANURAN POPULATIONS

HABITAT LOSS, DISTURBANCE, AND FRAGMENTATION

Habitat change poses the greatest threat to amphibians. A recent assessment identified agricultural activity as affecting 77 percent of species; timber and plant harvesting affecting 53 percent of species; and infrastructure development affecting 40 percent of species. As discussed in Chapter 8, maintaining well-connected suitable habitats within the landscape is therefore important for viable frog populations in the long term. (See the chart on page 252.)

CLIMATE CHANGE

As temperature and rainfall play fundamental roles in the lives of frogs, any changes in these climatic factors will clearly have far-reaching impacts. Climate change is already shifting the timing of breeding and patterns of distribution. Although species effects vary, in the Northern Hemisphere amphibians appear to be shifting toward earlier breeding faster than birds or butterflies. The frogs likely to be affected first are those living at the edge of their natural geographical range, as these peripheral populations may already be at their climatic limits. Such impacts will likely also influence individual growth and development.

INFECTIOUS DISEASE

Emerging infectious diseases can now explain some of the declines first observed in the 1990s. The viruses and fungi continuing to cause frog die-offs have probably been around for millennia, only recently becoming pathogenic as a result of frogs being transported around the world.

CHEMICAL CONTAMINANTS

The wide range of pesticides, herbicides, and fertilizers routinely sprayed on land can change rates of growth and development as well as alter behavior and physiology. They may also increase the vulnerability of frogs to other stressors such as disease. However, it is often difficult to disentangle the effects of chemicals from those of other aspects of habitat change associated with agriculture.

INVASIVE SPECIES

Introduced fish can be a major predator of frogs at different stages of their life cycle, as well as having indirect effects through influencing plants, invertebrates, and nutrient flow in ponds and streams. Equally, the introduction of mammalian predators, such as rats, cats, and mongooses, has had a devastating impact on many frog species, particularly on islands where native frogs lack natural defenses to novel invaders.

OVEREXPLOITATION

The trade in frogs can be divided into live trade for pets and laboratories, and animal products trade for food. Over 250 species of amphibians are threatened by over-exploitation for such purposes. Apart from the direct risks posed by over-collection, the transportation of frogs around the world for the live trade has been strongly linked to the spread of diseases. Frogs' legs may be traded using farmed or wild-caught frogs, but the demand around the world varies according to tradition and local cuisine.

DISEASES KILLING OFF FROGS

Ranaviruses were first detected in North America in the 1960s where they have been reported in ranids, bufonids, and hylids, as well as several species of salamander, some reptiles, and even fish. Ranavirus infections have also occurred in South America, Madagascar, Southeast Asia, and Australia. In mainland Europe they have caused die-offs of several species, including European Common Toads (*Bufo bufo*, see page 104), Common Midwife Toads (*Alytes obstetricans*), and European Common Frogs (*Rana temporaria*, see page 32) in southeast England in the 1980s and '90s (although some populations persist despite the infections). Chytridiomycosis is a disease of frogs caused by the fungus *Batrachochytrium dendrobatidis* (*Bd*). The fungus was first identified in captive South American poison frogs in 1997 but is thought to have originated from Southeast Asia before spreading through the shipment of amphibians around the world for the pet trade and laboratories. *Bd* has been implicated in the declines of over 500 amphibian species globally and may have caused more than 90 extinctions. This has included wiping out several species of harlequin frogs (genus *Atelopus*) as the disease spread through Central America. Fortunately, several of these species were taken into captivity before being hit by the disease. A further chytrid fungus—*Batrachochytrium salamandrivorans* (*Bsal*)—was discovered in 2013 in the Netherlands and Germany. Now known also to be a threat to frogs, *Bsal* has caused large-scale mortality of European salamanders and efforts are ongoing to reduce the risk of this disease spreading to other continents.

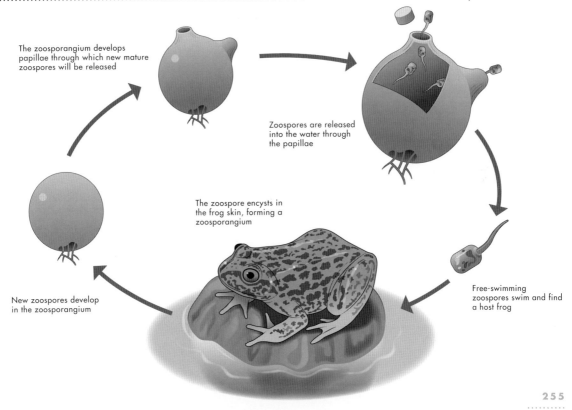

Life cycle of the chytrid fungus
Life cycle of the chytrid fungus (*Batrachochytrium dendrobatidis*) in the frog skin. The frog skin responds to infection by producing additional keratin, but this thickens it preventing metabolic processes (such as water uptake and respiration) and can lead to mortality from organ failure within days.

The zoosporangium develops papillae through which new mature zoospores will be released

Zoospores are released into the water through the papillae

The zoospore encysts in the frog skin, forming a zoosporangium

New zoospores develop in the zoosporangium

Free-swimming zoospores swim and find a host frog

Frog public relations

Ensuring that frogs have a future in a changing world is one of the most challenging conservation problems of our time. Public engagement is crucial for encouraging people to care about frogs and to highlight their importance to global biodiversity—doing so can feed into policies and actions to improve their protection. Scientists, conservationists, and just keen frog fans need to collaborate with individuals and organizations to deploy all the latest tools, technology, and ideas to help generate support for frogs.

History has a lot to answer for when it comes to present-day attitudes toward frogs. Things got off to a bad start in the 18th century when the father of modern taxonomy, Carolus Linnaeus, got it wrong with his otherwise ground-breaking *Systema Naturae*, in which he reiterated earlier classifications that placed frogs and other amphibians in the same group as reptiles, while also referring to them as "These foul and loathsome animals…" for good measure. This seems to have filtered through to today—a recent survey showed that 45 percent of respondents in local communities in South America had a strong aversion to the Ornate Horned Frog (*Ceratophrys ornata*). So, perhaps it is not surprising that frog conservation struggles for funding

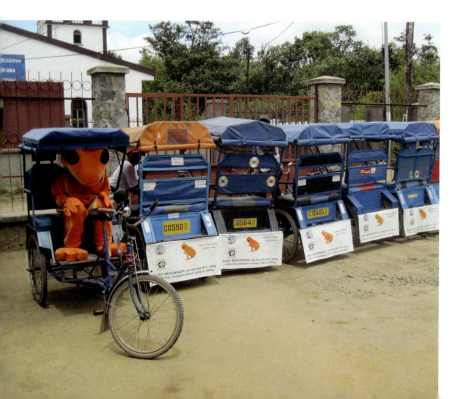

↗ Frogs feature on postage stamps throughout the world and some philatelists (stamp collectors) specialize in stamps featuring amphibians and reptiles.

← In Madagascar, the Golden Mantella (*Mantella aurantiaca*) is celebrated through motifs on the bicycles pedalled by tuc tuc drivers in the town of Moramanga.

→ In Yasothon, Thailand, the role of frogs in local culture is celebrated in a museum built to resemble a giant toad.

when up against charismatic megafauna like elephants, tigers, and gorillas (although we argue that the Ornate Horned Frog is indeed charismatic megafauna).

On the other hand, in some societies frogs are revered and used as cultural icons. In North America, frogs have been regarded as symbols of prosperity and healing, and today feature in traditional jewelry and art. In British Columbia frogs were incorporated on totem poles to ward off bad luck, and they have been used as symbols of fertility, prosperity, good luck, and, of course, transformation in cultures across the world. Accordingly, frog-themed ornaments and greetings cards are popular in gift stores everywhere.

WHAT CAN WE DO?

Improving the prospects of frogs can rely on effective frog PR. Whether or not frogs are already enshrined within local culture and history, developing narratives around their beauty and importance can change attitudes, increase knowledge, and ultimately encourage positive behaviors toward them. Practical, educational activities that bring people into contact with frogs work well, and indeed "Toads on Roads" campaigns have proved successful at securing anuran engagement. During breeding migrations, many thousands of toads (usually the European Common Toad, although other amphibian species are involved too) are rescued by volunteer groups across Europe as they attempt to cross busy roads en route to breeding ponds. Such exercises provide an ideal way of introducing the world of the toad to young and old alike, while also preventing toad-road mortality.

In 2008, the Amphibian Ark—a global coalition of zoos and aquaria—ran the "Year of the Frog" campaign, with the goal of raising awareness, increasing partnerships, carrying out conservation projects, and raising funds. A follow-up evaluation ten years later showed that the campaign had improved research and conservation as well as encouraging many institutions to embrace amphibian conservation.

Likewise, there are regional campaigns run by groups of enthusiasts across the world that can have significant impact at a local level. In the Mangabe region of Madagascar local groups have been established to produce clothing and tablecloths embroidered with images of the Critically Endangered Golden Mantella (*Mantella aurantiaca*). Sales of these products support the local community who also became guardians of the local breeding sites. Once such a groundswell of national, regional, or local pride has been established, the collective voice can begin to influence the decision-makers responsible for policy, legislation, and the protection of species and habitats.

Improving protection

Influencing those who make important decisions regarding the protection of species and their habitats can be challenging. A louder voice may be achieved by forging partnerships between organizations whose work happens to encompass the habitats used by frogs. Even when legal protection is in place, an absence of enforcement may mean a lack of compliance.

The legal frameworks regulating the protection of the environment and the species it contains are complex. There is enormous variation between different countries in how they do this, and even a species that is of national conservation importance may not be a priority for a regional government if it is abundant in that area. Various international legal instruments, such as the conventions that deal with biodiversity and climate change, have implications for frog conservation, but it depends on how these are translated into national policy, planning, and environmental protection frameworks.

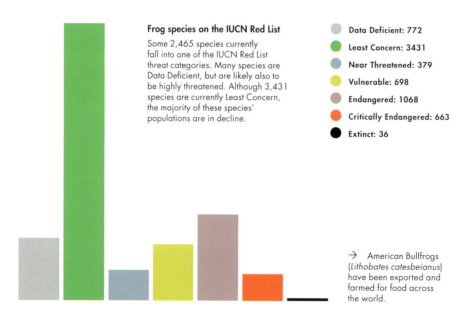

Frog species on the IUCN Red List
Some 2,465 species currently fall into one of the IUCN Red List threat categories. Many species are Data Deficient, but are likely also to be highly threatened. Although 3,431 species are currently Least Concern, the majority of these species' populations are in decline.

- Data Deficient: 772
- Least Concern: 3431
- Near Threatened: 379
- Vulnerable: 698
- Endangered: 1068
- Critically Endangered: 663
- Extinct: 36

→ American Bullfrogs (*Lithobates catesbeianus*) have been exported and farmed for food across the world.

IMPROVING PROTECTION

The USA has the Endangered Species Act of 1973—wherein threatened frog species and their habitats are protected—while the Lacey Act of 1900 prohibits illegal trafficking of certain species of animals and plants, and since 2016 has been effective in prohibiting the importation of salamanders that pose a risk of introducing *Bsal* to native amphibians. The European Union has several instruments, such as the Habitats Directive, under which member states are obliged to protect certain species from killing, disturbance, and habitat destruction. This includes over 20 species of frogs and certain important habitats where they occur.

Globally, the Convention on International Trade in Endangered Species (CITES) places obligations on its members to regulate trade in certain species that may be threatened by unsustainable collection in the wild, such as those popular for the pet trade (see the Harlequin Mantella, *Mantella cowanii*, page 270). However, at present only about 2 percent of amphibian species are CITES listed.

Enacting legislation to protect frogs and their habitats is in some ways the easy part of the process. The challenge is in making the legislation work, and especially so for small animals like frogs which may have hundreds of species in a given country, many either poorly known or perhaps not yet formally described. This means many frog habitats are easily destroyed simply because there are no data available on what species occur there; frogs are easily smuggled across borders in small containers without detection; and customs and border officials lack the expertise to identify frogs, so a rare protected species can easily be passed off as a common, legally exported species.

↑ Garden ponds can provide a haven for frogs and other wildlife.

↖ Guibé's Mantella (*Mantella nigricans*) is an attractive frog found only in Madagascar, where it occurs in several protected areas.

SOME SIGNS OF HOPE

Nevertheless, there have been positive trends for some species. Despite declines in the wider English countryside between the 1950s and 1970s, the increasing popularity of garden ponds in suburban and urban areas since has resulted in European Common Frogs (*Rana temporaria*, see page 32) moving in. In Brighton, on the south coast of England, it was estimated that one in seven urban gardens had a potential breeding site for amphibians. It is therefore remarkable how some pioneering frog species can find newly created ponds unassisted. Collectively, small ponds contain about two-thirds of all freshwater diversity, which in terms of size is proportionally more than lakes or rivers. Frog conservation can therefore be as simple as digging a hole and adding water, delivering a quick and easy way to improve the prospects of local frog populations as well as other fauna and flora.

Applying the science

Convincing policy makers, legislators, and indeed the wider public that frogs are important involves making compelling cases for their conservation. This requires strong arguments supported by good science. Although their cryptic lifestyles mean gathering important evidence to support frog conservation can be challenging, biologists and conservationists can employ novel tools to make their case.

Some 11 percent of amphibians on the IUCN's Red List are listed as "Data Deficient." This means there is insufficient information to assess their status or specific conservation needs. Before a species of frog can be protected it needs to be described and named, but if a species exists in a remote area and is behaviorally cryptic, this makes both discovering and describing it enormously challenging. In one area of Panama it is estimated that 30 species of frogs have been lost to disease, five of which are still yet to be formally described. The race is therefore on to get as many of the world's frogs described, so appropriate actions can be taken, and currently, new species of frogs are being described at the rate of approximately two per week.

Fortunately, rapid developments in science and technology are assisting frog researchers in finding and describing frogs and their habitats and identifying the best methods for their protection. Geographical Information Systems (GIS) can combine data on frog distribution patterns with sophisticated models of landscapes and climatic conditions to identify optimum areas for frogs and how to link fragmented habitats. Satellite imagery can allow habitats to be computationally reconstructed in three dimensions across vast areas, so future impacts of climate and habitat change can be estimated. The Hula Painted Frog (*Latonia nigriventer*) was thought to be extinct until its rediscovery in northern Israel in 2011. Since then, its distribution has been elucidated by searching for tiny amounts of its DNA (eDNA) left behind in water bodies (see pages 229 and 265). Combining these data with habitat information has enabled wider areas within the landscape to be identified where the frog may occur and need protection—all completed without the need to find a living frog.

← The Hula Painted Frog (*Latonia nigriventer*) was thought to be extinct until its rediscovery in Israel in 2011. Since then, its distribution has been revealed using environmental DNA (eDNA) sampling of wetlands.

APPLYING THE SCIENCE

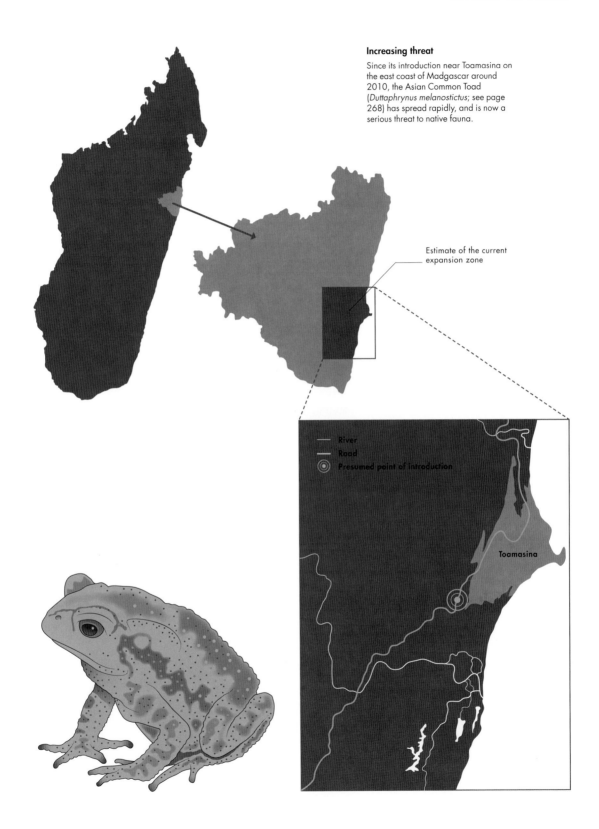

Increasing threat

Since its introduction near Toamasina on the east coast of Madagascar around 2010, the Asian Common Toad (*Duttaphrynus melanostictus*; see page 268) has spread rapidly, and is now a serious threat to native fauna.

FROGS IN THE FUTURE

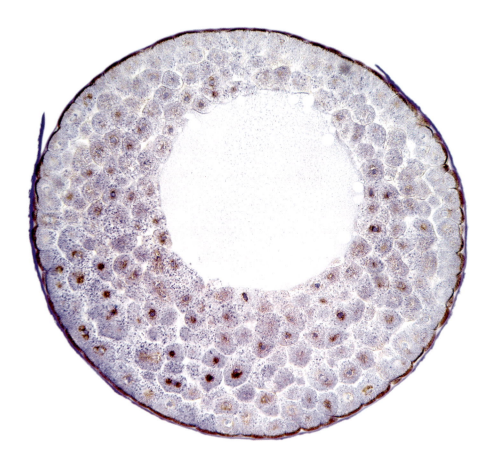

USING ADVANCES IN GENOMICS

Advances in understanding frog genetics through genomics has enormous potential to address a range of conservation issues. It is now possible to obtain the entire genetic code for given species, which can then be used alongside behavioral and experimental work to understand the genes or sections of the genome which code for, say, resistance to disease, meaning it may be possible to facilitate disease immunity or other adaptive properties in the future. Extracting ancient DNA (aDNA) from frogs collected and preserved in museums many years ago can now provide baseline genetic information against which present-day data on genetic diversity can be compared.

Genomics may also identify and help to recover small populations showing signs of inbreeding or declining genetic diversity. Although Genetically Modified Frogs (GMFs) raises a range of ethical issues and needs very careful scrutiny, it does have potential for helping to ensure a future for a range of species otherwise facing the prospect of extinction.

↑ A frog embryo showing cell division at the blastula stage. Artificial reproductive technologies using frozen eggs and sperm have the potential to substantially increase the numbers of frogs to assist conservation programs.

→ Surveys of the Otton Frog (*Babina subaspera*) in Japan have been carried out using eDNA sampling of breeding sites.

APPLYING THE SCIENCE

USING eDNA

Like all living organisms, frogs constantly shed tiny quantities of DNA into the environment. Samples of water, soil, and even air comprise a mixture of DNA from the animals and plants that either live in or have passed through a given habitat. Molecular methods are now available to extract, amplify, and identify the environmental DNA (eDNA) of those organisms—from just a small sample of pond water, for example. A presence/absence detection study on three species of Japanese frogs—the Otton Frog (*Babina subaspera*), Amami Tip-nosed Frog (*Odorrana amamiensis*), and Amami Oshima Frog (*O. splendida*)— illustrated a mismatch between acoustic survey methods and eDNA, in that acoustic surveys usually only detect calling males, while eDNA can detect males, females, and tadpoles. Although ecology, life stage, and seasonality can influence eDNA detections, the recent addition of this method as a tool for monitoring frog populations is facilitating research on a scale not previously possible.

CAPTIVE BREEDING AND ASSISTED REPRODUCTION

The amphibian decline crisis has generated renewed interest in using captive breeding as a conservation tool. Reintroduction of captive bred frogs into the wild was originally the main driver for breeding them in zoos, aquaria, and research institutions, and examples include the Kihansi Spray Toad (*Nectophrynoides asperginis*, see page 272). Indeed, there are many examples of such reintroductions being successful. It is estimated that about 25 percent of Mallorcan Midwife Toads (*Alytes muletensis*, see page 102) in the wild have resulted from the release of captive bred stock.

However, the role of captive colonies has broadened beyond simply providing animals for release into the wild. In addition to the opportunity to spread the word about the amphibian crisis to a global audience of many millions of zoo visitors, cutting-edge conservation research is being conducted on captive frogs that would not be possible with wild populations. Much of our understanding of how frogs may respond to disease-causing fungi has been gained through trials involving captive animals. Indeed, some bacteria that naturally live on the skin of frogs can mop up disease-causing fungi, and this is providing insights into how disease might be mitigated in the wild (for example, Australia's Southern Corroboree Frog, *Pseudophryne corroboree*—see page 276).

Assisted Reproductive Technologies (ART) are helping to resolve challenges around keeping and breeding frogs in captivity. Rather than maintaining breeding colonies of multiple species, samples of their genetic material—such as sperm, eggs, and tissue—can be obtained and stored, and then used selectively to address declining genetic diversity and increase adaptive potential. In this way, frog populations could be better equipped to adapt to the increasing ecological and environmental threats they face in the wild. For example, artificial fertilization of frog eggs has proved successful in several species, and in at least three—the Oregon Spotted Frog (*Rana pretiosa*), and the White-bellied (*Geocrinia alba*) and Orange-bellied Frogs (*G. vitellina*) from Australia—the application of "biobanking" substantially reduces the numbers of frogs needed to maintain a captive breeding colony and reduces inbreeding in frogs produced by the program.

→ Several zoos in North America are collaborating on research for breeding the rare Oregon Spotted Frog (*Rana pretiosa*) as part of its species recovery program.

FROGS IN THE FUTURE

DUTTAPHRYNUS MELANOSTICTUS

Asian Common Toad

One of the few anurans with an increasing population

SCIENTIFIC NAME:	*Duttaphrynus melanostictus*
FAMILY:	Bufonidae
LENGTH:	2–6 in (57–150 mm)
LIFE HISTORY:	Aquatic eggs (up to 10,000 in strings) and tadpoles; terrestrial adults
NOTABLE FEATURE:	Highly adaptable and tolerant of disturbed habitat
IUCN RED LIST:	Least Concern

The Asian Common Toad has the typical appearance of a toad, with a large body (some females may reach 8 in/20 cm), short sturdy legs, and warty skin. Its general coloration is highly variable and ranges from gray to brown, orange, and yellow, with mottled patterning and sometimes with reddish patches. Its most characteristic feature is the black/dark brown ridged crest which runs around and between the eyes to the nose, combined with a dark edge to the mouth and dotted dark spots across the body

Several life-history features—also shared with the highly invasive Cane Toad (*Rhinella marina*)—make the Asian Common Toad such an adaptable anuran. These include a robust physiology that allows them to survive in dry, exposed habitat; short, sturdy legs suited to overground dispersal; an ability to breed in permanent, ephemeral, or slow-moving bodies of water; high fecundity; fast tadpole development; a long life span; high tolerance of disturbance; and strong predatory defenses.

In Madagascar, a breeding population of Asian Common Toads was discovered in 2014 around the port of Toamasina on the east coast (see page 263). Thought to have originated from toads accidentally imported with construction equipment, the population increased so quickly that within three years of its discovery the toads had dispersed more than 12 miles (20 km) from the original introduction site, rendering an eradication program impractical.

As the Asian Common Toad has skin toxins to which the endemic Madagascan fauna likely have no natural resistance, the fear is that when they expand into protected forest areas this could have devastating impacts. Equally, there is a risk that Asian toads could spread novel pathogens into native frog populations. As a result, current efforts are focusing on trying to slow the ongoing dispersal of Asian Common Toads through public education, improved biosecurity, and implementing barriers to dispersal.

In its native range, when preying on the Asian Common Toad, the Kukri Snake (*Oligodon fasciolatus*) will use its teeth to cut into the abdomen of the toad, whereupon the snake eats the toad's internal organs! Some birds have been observed to feed on bufonid toads in a similar way, and this is likely to avoid ingesting defensive toxins.

→ Widely distributed in South and Southeast Asia, the Asian Common Toad may in fact comprise a complex of multiple species, some of which have been introduced to islands such as Bali, Sulawesi, New Guinea, and more recently Australia and Madagascar, where there are concerns about impacts on native fauna.

FROGS IN THE FUTURE

MANTELLA COWANII

Harlequin Mantella

One of the most threatened Malagasy frogs

SCIENTIFIC NAME:	*Mantella cowanii*
FAMILY:	Mantellidae
LENGTH:	¾–1¼ in (20–30 mm)
LIFE HISTORY:	Unconfirmed, but likely similar to other *Mantella* species (terrestrial eggs and adults; aquatic tadpoles)
NOTABLE FEATURE:	Adapted its behavior to survive in degraded habitat
IUCN RED LIST:	Endangered

Confined to just four known locations in central Madagascar—none of which fall within a protected zone—and occupying an area of less than 4 square miles (10 square kilometers), there may be fewer than 200 Harlequin Mantellas left in the wild. Deforestation and habitat fragmentation have resulted in the remaining populations becoming isolated, while the frog's attractive color pattern means it has also been heavily collected for the pet trade.

Historically widely distributed across continuous upland forest, these habitats have been all but lost, leaving the largest remaining wild population of Harlequin Mantella to occupy degraded (now savannah) habitat, where it is confined to damp seepages on rocky slopes and a subterranean stream. In this area, Harlequin Mantellas are only active early in the morning and late in the afternoon, as there is no remaining forest cover for the frogs during the heat of the day. The strikingly marked black body—with orange-red markings on the legs, groin, and armpits, and round bluish spots on the underside—make this species highly sought after in the international pet trade. It is estimated that over 3,000 Harlequin Mantellas were legally exported from Madagascar between 1994 and 2003, after which trade was halted.

Although now fully protected by law, illegal collection for the trade remains a threat, along with habitat disturbance and fires.

A further challenge to the persistence of the Harlequin Mantella comes from hybridization with another species, the similarly patterned Baron's Mantella (*Mantella baroni*). At least one subpopulation of the Harlequin Mantella exists in sympatry with Baron's Mantella, where they are known to hybridize. Interbreeding produces color morphs that are intermediate between the two species, and a significant percentage of the Harlequin Mantella population in this area is now thought to be of hybrid origin. Other subpopulations of the Harlequin Mantella found at higher elevations are also at risk, as climate change may promote an upward range shift in the seemingly more adaptable Baron's Mantella.

A species action plan has been implemented for the Harlequin Mantella, focused on habitat protection, field research, support for local communities, environmental awareness, training, and information sharing.

→ The striking red coloration on the black body and legs of the Harlequin Mantella contrasts with bold blue spots on the underside. The patterns of these markings are unique to each frog, and have been used by researchers to identify individual frogs in the remaining populations.

FROGS IN THE FUTURE

NECTOPHRYNOIDES ASPERGINIS

Kihansi Spray Toad

The waterfall-spray specialist

SCIENTIFIC NAME:	*Nectophrynoides asperginis*
FAMILY:	Bufonidae
LENGTH:	½–1 in (10–25 mm)
LIFE HISTORY:	Up to 13 tadpoles retained within the female, direct developing, and live birth of froglets
NOTABLE FEATURE:	Translucent ventral skin shows internal organs and (in females) developing young
IUCN RED LIST:	Extinct in the Wild

Discovered in 1996, the Kihansi Spray Toad was only found in a tiny area of Tanzania, where it lived in the spray zone of torrents that descended the Kihansi Gorge. The diversion of the water to generate hydroelectricity resulted in drier habitats and a dramatic decline in numbers. To rectify this, artificial sprinkler systems were installed in 2000–2001 to rehydrate the wetlands. As a backup plan, several hundred toads were collected for captive breeding in North American zoos.

Initially, the sprinkler systems resulted in much improved habitats for the Kihansi Spray Toad, but in 2003 the species went into a further sharp decline, and in 2009 it was declared Extinct in the Wild by IUCN. Habitat change and poor water quality may have contributed to the extinction, but the primary driver is now known to have been the chytrid fungus (*Bd*). This may have been introduced through the accidental movement of infected frogs from other areas along with building materials associated with the dam construction. Fortunately, toads in North American zoos bred successfully, and some were returned to Tanzania where pilot reintroductions were carried out between 2013 and 2017. Subsequent surveys revealed very few toads, indicating that disease and possibly other threats remain an issue for this species.

→ Along with the Wyoming Toad (*Anaxyrus baxteri*), the Kihansi Spray Toad is currently classified as Extinct in the Wild on the IUCN Red List. This is because the remaining in situ populations rely on captive-bred animals to support their "wild" population.

FROGS IN THE FUTURE

RANA DALMATINA

Agile Frog
Long-legged leaper

SCIENTIFIC NAME:	*Rana dalmatina*
FAMILY:	Ranidae
LENGTH:	2½–3½ in (60–85 mm)
LIFE HISTORY:	Aquatic eggs (250–1,800 in clumps) and tadpoles; terrestrial adults
NOTABLE FEATURE:	Adult frogs can jump a distance of up to 6 ft (2 m) at a time
IUCN RED LIST:	Least Concern

The Agile Frog is remarkably similar to the European Common Frog, but it can be distinguished by its pointed snout, pale unspotted underside, and exceptionally long back legs. There is also sometimes a yellowish patch around the base of the thighs. It is widespread in southeastern, central, and western Europe, and is the only native frog on the Channel Island of Jersey, where it is near the edge of its range. Although it is a frog of deciduous woodlands and wet meadows in mainland Europe, in Jersey it is found on coastal heathlands, where it breeds in shallow, temporary pools that fill during the winter and usually dry up in the summer.

As a result of disturbance and loss of habitat on Jersey, Agile Frogs went into decline on the island in the early 20th century. Factors, including pollution and the introduction of predators reduced the number of known breeding sites to just seven by the 1970s. By the 1990s, only a single breeding locality remained, and a Species Action Plan was initiated in 2001 by a group of local partners, including the States of Jersey Government. A component of the plan was to protect spawn clumps from aquatic predators by surrounding them with nets during their development, and to collect some spawn to hatch and raise in captivity at Jersey Zoo.

In natural conditions on Jersey, Agile Frogs lay small clumps containing up to 250 eggs attached to a submerged branch or stem. Eggs then hatch into tadpoles, which take 2–3 months to metamorphose, with sexually mature adults returning to breed 2–3 years later. Using a conservation technique known as headstarting to aid recruitment of Jersey's Agile Frog population, egg clumps were collected and tadpoles raised in captivity until they were large enough to avoid predators, at which time these advanced stage tadpoles were returned to the breeding ponds. By 2017 some 167 spawn clumps had been protected in the ponds, and nearly 200 spawn clumps had been headstarted at the zoo, resulting in over 57,000 tadpoles being released at four sites on the island. Combined with other conservation efforts over more than three decades, this has resulted in an increase in natural spawning from fewer than 20 clumps in 1987 to over 150 by 2015.

→ The Agile Frog is an explosive aquatic breeder, and females lay individual clumps of spawn. Counting these can provide an indication of the number of females that have bred in a given pond and year, and aid in understanding how many individuals may be actively reproducing in a population, known as the effective population size.

FROGS IN THE FUTURE

PSEUDOPHRYNE CORROBOREE

Southern Corroboree Frog

Captive-breeding, disease-resistant frogs may ensure species survival

SCIENTIFIC NAME:	*Pseudophryne corroboree*
FAMILY:	Ranidae
LENGTH	Myobatrachidae
LIFE HISTORY	Terrestrial eggs (10–38 in deep burrows), aquatic tadpoles, terrestrial adults; parental care
NOTABLE FEATURE	Strikingly marked with broken yellow, longitudinal stripes on a black body
IUCN RED LIST	Critically Endangered

Now recognized as distinct from the Northern Corroboree Frog (*Pseudophryne pengilleyi*), the distributions of both species have contracted to two small upland areas close to Canberra in southeastern Australia. The remaining Southern Corroboree Frogs may now number less than 50 individuals in the wild in Kosciuszko National Park: this means the population is probably no longer viable and the species is functionally extinct in the wild. Fortunately, Corroboree Frogs are well established in a several zoos and research institutes where vital research to ensure the future of both species is being carried out.

One of the main drivers of decline of the species is the chytrid fungus (*Bd*). Although it is theoretically possible, reintroduction of Southern Corroboree Frogs to their former range is problematic as *Bd* persists in other frog species in the same area, yet these frogs do not succumb to the disease.

However, recent research suggests Corroboree Frogs hold sufficient genetic variation to evolve some degree of disease resistance. Consequently, applying novel genetic technologies has the potential to develop strains of disease-resistant frogs that can ultimately be released safely in the wild. The project suffered a setback in 2020 when bushfires swept through some disease-proof enclosures containing Corroboree Frogs in Kosciuszko National Park. Remarkably, more than 30 percent of the frogs survived the fire by seeking refuge underground, and the enclosures have since been reconstructed to allow the conservation research to continue.

→ The diet of Corroboree Frogs can influence the presence or absence of certain bacteria on their skin. This has implications for the wider naturally occurring skin microbial community (microbiome) of these frogs, as such communities can determine how resistant an individual is to disease such as chytridiomycosis, caused by the chytrid fungus.

GLOSSARY

↑ Modern frogs—like this stunning rhacophorid treefrog—have been around for more than 120 million years and are among the most adaptable and successful vertebrates on the planet. With so much yet to discover about their fascinating lives and natural histories, we hope you agree they truly are intriguing and inspiring animals, and that this book has piqued your interest to find out more.

Aggregation Clustering of individuals due to (for example) reproductive activity, feeding, or defense.

Amphibian Vertebrate animal with soft permeable skin and a typically bi-phasic (aquatic followed by terrestrial) life cycle.

Amplexus The anuran mating position in which males grasp a female in advance of and during egg laying.

Anura The taxonomic order to which all tailless amphibians (frogs and toads) belong.

Aposematism The use of distinctive coloration and patterning to advertise the toxicity of an animal to potential predators.

Arboreal Living in or among trees.

Biodiversity The variety of living organisms at all scales, from individual genes to whole ecosystems.

Brumation A state of inactivity or torpor in response to cold conditions.

Buccal cavity The oral cavity (see also buccopharyngeal cavity).

Buccopharyngeal cavity The entirety of the oral cavity including the pharynx.

Chromosomes Structures in the nucleus of a cell that carry the genetic material.

Clade A group of organisms comprising a common ancestor and all its descendants.

Cloaca The common chamber into which reproductive (sperm, eggs) and excretory ducts (waste) open.

Convergent evolution Evolution of similar characteristics in unrelated organisms.

Cryoprotectorant A naturally occurring chemical that prevents freezing.

Cryptic Secretive or difficult to observe on account of camouflage or behavior.

Direct development Completion of development within the egg, whereupon hatching a froglet emerges.

Dormancy Any period of inactivity when growth and development stops.

Ectotherm An animal in which body temperature is regulated through external heat and associated behavior (behavioral thermoregulation).

Endotherm An animal in which body temperature is regulated internally and at a different level to the ambient temperature.

Estivation A continuous state of inactivity during hot and dry periods.

Foam nest A protective environment generated by some breeding frogs into which eggs and sperm are released, and eggs and tadpoles can then develop.

Founder effect Reduction in genetic variation that is a consequence of a small number of individuals founding a new biologically separate population.

Froglet A recently metamorphosed or young frog.

Frugivory Eating of fruit.

Gastromyzophory The condition in tadpoles in which the belly is modified into an abdominal sucker.

Genetic rescue Introduction of novel genetic material to a population which boosts genetic diversity.

Genetics The study of genes and heredity.

Genomics The study of all the DNA contained within an organism (i.e., the genome).

Geographic distribution The complete range over which a species is found in nature.

Hibernation A continuous state of inactivity during cold periods.

Hybridization When two species interbreed and produce offspring.

Hyperossified Presence of a greater amount of bone material, extensively ossified (see also ossification).

Inbreeding Mating between closely related individuals.

Inducible defense A protective or behavioral response that arises from a threat.

Lateral line organ A pressure-sensitive organ (also known as a neuromast) of which many are usually arranged in specific patterns on the body forming the lateral line system.

Lek A temporary defended area selected by (usually) males in which to courtship display.

Lentic Living in still freshwater.

Metamorphosis (plus metamorph) The transformation from a tadpole into a froglet; a recently metamorphosed froglet may be known as a metamorph.

Morphology The physical features that comprise the form and structure of an organism.

Myrmecophagy Consumption of ants and/or termites.

Necrophagy Consumption of dead animals (carrion).

Obligate oophagy When an animal is exclusively dependent on (normally unfertilized) eggs for food.

Opercularis system (see also operculum) An additional auditory pathway unique to anurans comprising the shoulder/forelimb skeleton and muscles that connect to the inner ear.

GLOSSARY

Operculum A structure that covers or closes an opening. In frogs, this can relate to both the flap of skin which grows over and protects the gills of a tadpole, and the bony structure that sits within the inner ear.

Ossification The process of bone formation (*see also* hyperossified).

Osteoderm Bony scale or plate embedded in the skin.

Oviparity Laying eggs.

Ovoviviparity Retention and hatching of eggs within the female oviduct, such that fully developed young are live birthed.

Parental care Investment by one or both parents in protecting their offspring in some way (for example, attending the eggs, internal brooding).

Parotoid gland A toxin-containing glandular region on/behind the head.

Population bottleneck A rapid and significant reduction in population size which causes a considerable loss in genetic diversity.

Refugia A place of refuge used by an animal to escape (for example) predators or environmental conditions (*see also* refugium).

Refugium (plural: refugia) A geographical location to which a species may be restricted as a result of widespread climatic change (during glaciation for example).

Riparian Living alongside a river or stream.

Sink A population where deaths or emigration are higher than births or immigration (*see also* source).

Skeletochronology Technique used to estimate the age of a vertebrates by counting lines of arrested growth within skeletal tissues.

Sneaky mating (sexual parasitism) Typically in males, an alternative reproductive strategy whereby an individual will exploit the efforts of a displaying male to intercept and mate with a female attracted by the displaying male.

Source A population where births or immigration are higher than deaths or emigration.

Speciose A group or taxon that contains many species.

Spiracle Opening through which water passes after flowing over the gills of tadpoles.

Sympatry When organisms occur within the same geographical area.

Taxon (plural: taxa) A group of taxonomically related organisms (for example, family, genus, or species).

Thanatosis Death feigning, usually in response to a predator.

Transitional fossil A fossil that has characteristics of both ancestral (for example, extinct) and derived (for example, extant) forms.

Tubercle A small, rounded protuberance on the skin.

Tympanum (singular: tympani) The eardrum or tympanic membrane. Visible in many frogs externally as a circular membrane behind the eye.

Unken reflex Extreme arching of the body in response to a threat from a predator, often to reveal warning coloration.

Viviparity Development of young within the female through direct provision of nutrients before live birth.

Vocal sac Highly elastic skin membrane used to resonate and project advertisement calls.

Vocalization The audible calls produced by an animal.

RESOURCES

BOOKS

Carroll, R. L. *The Rise of Amphibians: 365 Million Years of Evolution* (Baltimore: Johns Hopkins University Press, 2009)

Dodd, C. K. Jr. (Ed.) *Amphibian Ecology and Conservation: A Handbook of Techniques* (Oxford: Oxford University Press, 2010)

Dodd, C. K. Jr. *Frogs of the United States and Canada* (Baltimore: Johns Hopkins University Press, 2023)

Duellman W. E., and L. Trueb. *Biology of Amphibians* (New York: McGraw-Hill, 1994)

Elliott, L., Gerhardt C., and C. Davidson. *Frogs and Toads of North America* (North America: Houghton Mifflin Harcourt, 2009)

Gerhardt, H. C., and F. Huber. *Acoustic Communication in Insects and Anurans: Common Problems and Diverse Solutions* (London: University of Chicago Press, 2002)

Lannoo, M. (Ed.) *Amphibian Declines: The Conservation Status of United States Species* (London: University of California Press, 2005)

McDiarmid, R. W., and R. Altig (Eds.). *Tadpoles: The Biology of Anuran Larvae* (London: University of Chicago Press, 1999)

O'Shea, M. and Maddock, S. *Frogs of the World: A Guide to Every Family* (Oxford: Princeton University Press, 2024).

Pough, F. H., Andrews, R. M., Crump, M. L., Savitzky, A. H., Wells, K. D., and M. C. Brandley. *Herpetology*. 4th edition. (Sunderland Massachusetts: Sinauer Associates Incorporated, 2014)

Pough, F. H., Janis, C. M., and J. B. Heiser. *Vertebrate Life*. 9th edition. (London: Pearson, 2013)

Ryan M. (Ed.). *Anuran Communication* (London: Smithsonian, 2001)

Schoch, R. R. *Amphibian Evolution: The Life of Early Land Vertebrates* (Chichester: John Wiley and Sons, 2014)

Vitt L. J., and J. P. Caldwell. *Herpetology: An Introductory Biology of Amphibians and Reptiles*. 3rd edition. (London: Academic Press, 2014)

Wells, K. D. *The Ecology and Behavior of Amphibians* (London: The University of Chicago Press, 2007)

SELECTED SCIENTIFIC JOURNAL ARTICLES

Barrile, G. M., et al. "Wildfire influences individual growth and breeding dispersal, but not survival and recruitment in a montane amphibian." *Ecosphere*, 13 (8): e4212 (2022). https://doi.org/10.1002/ecs2.4212

Beebee, T. J. C. "Genetic contributions to herpetofauna conservation in the British Isles." *Herpetological Journal*, 28 (2): 51–62 (2018)

Chambert, T., et al. "Defining relevant conservation targets for the endangered Southern California distinct population segment of the mountain yellow-legged frog (*Rana muscosa*)." *Conservation Science and Practice*, 4 (5): e12666 (2022). https://doi.org/10.1111/csp2.12666

Claas B. and J. Dean. "Prey-capture in the African clawed toad (*Xenopus laevis*): comparison of turning to visual and lateral line stimuli." *Journal of Comparative Physiology* A 192: 1021–1036 (2006)

Daeschler E. B., et al. "A Devonian tetrapod-like fish and the evolution of the tetrapod body plan." *Nature*, 440 (7085): 757–63 (2006). https://doi.org/10.1038/nature04639

de-Oliveira-Nogueira, C. H., et al. "Between fruits, flowers and nectar: The extraordinary diet of the frog *Xenohyla truncata*." *Food Webs*, 35: e00281 (2023). https://doi.org/10.1016/j.fooweb.2023.e00281

Fischer, E. K. "Form, function, foam: evolutionary ecology of anuran nests and nesting behaviour." *Philosophical Transactions of the Royal Society* B, 378 (1884): 20220141 (2023). https://doi.org/10.1098/rstb.2022.0141

Grafe, T. U., et al. "Demographic dynamics of the afro-tropical pig-nosed frog, *Hemisus marmoratus*: effects of climate and predation on survival and recruitment." *Oecologia*, 141: 40–46 (2004). https://doi.org/10.1007/s00442-004-1639-7

Guayasamin, J. M., et al. "Phenotypic plasticity raises questions for taxonomically important traits: a remarkable new Andean rainfrog (*Pristimantis*) with the ability to change skin texture." *Zoological Journal of the Linnean Society*, 173 (4): 913–928 (2015). https://doi.org/10.1111/zoj.12222

Hime P. M. et al. "Phylogenomics reveals ancient gene tree discordance in the amphibian Tree of Life." *Systematic Biology*, 70 (1): 49–66 (2021). https://doi.org/10.1093/sysbio/syaa034

Hurley S. L. "The amphibians of Monte Alén National Park: bioacoustics, ecology and conservation." University of Bristol, MSc Thesis (2023)

Jared, C., et al. "Venomous frogs use heads as weapons." *Current Biology*, 25(16): 2166–2170 (2015). http://dx.doi.org/10.1016/j.cub.2015.06.061

Lamoureux, V. S., et al. "Premigratory autumn foraging forays in the green frog, *Rana clamitans*." *Journal of Herpetology*, 36 (2): 245–254 (2002). https://doi.org/10.1670/0022-1511(2002)036[0245:PAFFIT]2.0.CO;2

Licata, F., et al. "Abundance, distribution and spread of the invasive Asian toad *Duttaphrynus melanostictus* in eastern Madagascar." *Biological Invasions*, 21 (5): 1615–1626 (2019). https://doi.org/10.1007/s10530-019-01920-2

Liedtke, H. C., et al. "The evolution of reproductive modes and life cycles in amphibians." *Nature communications*, 13 (1): 7039 (2022). https://doi.org/10.1038/s41467-022-34474-4

Muths, E., et al. "First estimates of the probability of survival in a small-bodied, high-elevation frog (boreal chorus frog, *Pseudacris maculata*), or how historical data can be useful." *Canadian Journal of Zoology*, 94 (9): 599–606 (2016). https://doi.org/10.1139/cjz-2016-0024

Pašukonis, A., et al. "Map-like navigation from distances exceeding routine movements in the three-striped poison frog (*Ameerega trivittata*)." *Journal of Experimental Biology*, 221 (2): jeb169714 (2018). https://doi.org/10.1242/jeb.169714

Pechmann, J.H., et al. "Declining amphibian populations: the problem of separating human impacts from natural fluctuations." *Science*, 253 (5022): 892–895 (1991). https://doi.org/10.1126/science.253.5022.892

Rojas Zuluaga, B. "Behavioural, ecological, and evolutionary aspects of diversity in frog colour patterns". *Biological Reviews*, 92 (2): 1059–1080 (2017). https://doi.org/10.1111/brv.12269

Sinsch, U. "Movement ecology of amphibians: from individual migratory behaviour to spatially structured populations in heterogeneous landscapes." *Canadian Journal of Zoology*, 92 (6): 491–502 (2014). https://doi.org/10.1139/cjz-2013-0028

Taboada, C., et al. "Glassfrogs conceal blood in their liver to maintain transparency." *Science*, 378 (6626): 1315–1320 (2022). https://doi.org/10.1126/science.abl6620

Toledo, L. F., et al. "Behavioural defences of anurans: an overview." *Ethology Ecology and Evolution*, 23 (1): 1–25 (2011). https://doi.org/10.1080/03949370.2010.534321

Vági, B., et al. "Parental care and the evolution of terrestriality in frogs." *Proceedings of the Royal Society B*, 286 (1900): 20182737 (2019). https://doi.org/10.1098/rspb.2018.2737

Warkentin, K. M. "Plasticity of hatching in amphibians: evolution, trade-offs, cues and mechanisms." *Integrative and Comparative Biology*, 51 (1):111–127 (2011). https://doi.org/10.1093/icb/icr046

Weldon, C., et al. "Disease driven extinction in the wild of the Kihansi spray toad, *Nectophrynoides asperginis*." *African Journal of Herpetology*, 69 (2): 151–164 (2021). https://doi.org/10.1139/cjz-2016-0024

Yuan, M.L., et al. "Endemism, invasion, and overseas dispersal: the phylogeographic history of the Lesser Antillean frog, *Eleutherodactylus johnstonei*." *Biological Invasions*, 24 (9): 2707–2722 (2022). https://doi.org/10.1007/s10530-022-02803-9

AMPHIBIAN ORGANIZATIONS, HERPETOLOGY SOCIETIES, AND USEFUL WEBSITES

Amphibian and Reptile Conservation (https://www.arc-trust.org)

Amphibian Ark (https://www.amphibianark.org)

Amphibian Species of the World (https://www.amphibiansoftheworld.amnh.org)

Amphibian Survival Alliance (https://www.amphibians.org)

AmphibiaWeb (https://www.amphibiaweb.org)

British Herpetological Society (https://www.thebhs.org)

Herpetological Society of Africa (https://www.africanherpetology.org)

Herpetological Society of Japan (https://www.herpetology.jp)

Herpetological Society of Singapore (https://www.herpsocsg.com)

Herpetologists' League (https://www.herpetologistsleague.org)

International Herpetological Society (https://www.ihs-web.org.uk)

IUCN SSC Amphibian Specialist Group (https://www.iucn-amphibians.org)

IUCN Red List of Threatened Species (https://www.iucnredlist.org)

Sociedade Brasileira de Herpetologia (https://www.sbherpetologia.org.br)

Societas Europaea Herpetologica (https://www.seh-herpetology.org)

Society for the Study of Amphibians and Reptiles (https://www.ssarherps.org)

Women in Herpetology (https://www.womeninherpetology.com)

World Congress of Herpetology (https://www.worldcongressofherpetology.org)

INDEX

Acanthixalus spinosus 198
Acris crepitans 76, 148
Acris gryllus 148
African Bullfrog 84, 119, 130–1
African Clawed Frog 13, 52, 122, 124, 178
African reed frogs 30
African Spiny Frog 198
Agalychnis callidryas 197, 208, 210–11
Agile Frog 237, 274–5
alkaloids 202
Allobates femoralis 150
Alytes muletensis 88, 102–3, 197, 231, 266
Alytes obstetricans 255
Amami Oshima Frog 265
Amami Tip-nosed Frog 265
ambush predators 174–5
Ameerega trivittata 150
American Bullfrog 95, 96–7, 152, 200, 250
American Green Treefrog 47, 64–5, 88
American Toad 34–5, 123, 196, 253
Amolops 123
Amolops larutensis 132–3
Amphibian Ark 257
amplexus 89, 94, 126
Anaxyrus americanus 34–5, 123, 196, 253
Anaxyrus boreas 145, 162–3
Anaxyrus fowleri 225
Anaxyrus terrestris 170, 176
ancient DNA (aDNA) 264
Andes Smooth Frog 53
aposematism 98, 156, 175, 193, 198–9, 200, 204–5, 208, 218
Aquarana heckscheri 170
arboreality 46–7, 143
   gliding 47, 158
   see also treefrogs
Archaeobatrachia 15–16, 28, 177, 179
Archey's Frog 36–7

ascaphid (tailed) frogs 26, 38–9, 94, 140
Ascaphus montanus 38–9
Asian Common Toad 250, 268–9
Asian treefrogs 47
assisted reproductive technologies (ARTs) 242, 266
Atelopus 255
Atelopus zeteki 83, 98–9
Atympanophrys gigantica 123
Aubria subsigillata 170

Babina subaspera 265
backpack nurseries 113
Bahia Green Treefrog 198
Barbourula kalimantanensis 54
Baron's Mantella 270
Batesian mimicry 209
Batrachochytrium see chytrid fungus
beaconing 146
biobanking 266
biting 199
Blommersia transmarina 30
Boana 84
Bolivian Bleating Frog 186
Bombina 199
Boophis madagascariensis 80
Boophis nauticus 30
Boreal Chorus Frog 224, 238–9
Bornean Black-spotted Rock Frog 82
Bornean Flat-headed Frog 54
Borneo Narrow-mouthed Frog 45
Brachycephalus pitanga 140, 156–7
breeding *see* reproduction
breeding chorus 55, 93, 95
Breviceps 94
Brilliant-thighed Poison Frog 150
bromeliads 44–5, 114, 117, 128, 143
brood chambers 118

West African Brown Frog 170
Budgett's Frog 122, 182–3
Bufo bufo 30, 92, 104–5, 114, 124, 143, 148, 200, 233, 255
Bufotes viridis 145
Bumblebee Poison Frog 208

Callobatrachus sanyanensis 23
camouflage 32, 82, 158, 175, 186, 193–4, 204–5, 214
Cane Toad 149, 152, 170, 176, 250
cannibalism 168, 170
captive breeding 102, 242, 244, 266, 272, 274
capture-mark-recapture 235
carnivores 168
   see also predation by frogs
carotenoids 158
carrion 53, 168, 170
ceratophrid frogs 175, 182, 186–7, 199, 256–7
Ceratophrys 199
Ceratophrys cornuta 175, 186–7
Ceratophrys ornata 256–7
chemical contaminants 254
chemical cues 77, 212
chemical defenses 200–3
   see also toxins
Chicxulub asteroid 27
Chile Darwin's Frog 117
chorus frogs 60, 238–9
chromatophores 194, 205
chytrid fungus 66, 98, 152, 153, 203, 255, 272, 276
classification 14–17
clawed frogs 13, 52, 53, 122, 124, 178, 179, 188–9
climate change 40, 232–3, 252, 254
co-ossification 156
cocoons 70
cognitive maps 150
colonization 144–5
coloration
   warning 98, 156, 175, 193,

198–9, 200, 204–5, 208, 218
   see also camouflage; skin
Colostethus 94
Columbia Spotted Frog 231
Common Midwife Toad 255
communication 76–88
compass orientation 146–8
Concave-eared Torrent Frog 55
Congo Dwarf Clawed Frog 178, 188–9
Conraua 77
Conraua goliath 31
conservation 36, 232–3, 242, 244, 246, 256–67, 270–6
Convention on International Trade in Endangered Species (CITES) 260
convergent evolution 27, 46, 59, 171, 201
Coquí Frog 58
Couch's Spadefoot Toad 59, 68–9
cricket frogs 148
Crossodactylodes itambe 45, 143
Crowned Poison Frog 218
Cruziohyla 46
Cruziohyla calcarifer 46
Cruziohyla craspedopus 46
Cruziohyla sylviae 46
cultural icons 257
Cururu Toad 170, 173
Cyclorana platycephala 60

Darwin, Charles 86, 117
Darwin's Frog 117
death-feigning 198
Dendrobates 94
Dendrobates auratus 150
Dendrobates leucomelas 208
dendrobatid frogs 94, 128, 141, 150, 175, 201, 202, 208
desiccation risk 21, 50–1, 58–9
Devonian Period 20–1
direct development 31, 58, 111, 112, 250
disease *see* pathogens

284

INDEX

dispersal 30, 45, 56, 64, 123, 141, 144–5, 152–4, 162, 228–33
   barriers 231, 233
   Marbled Snout-burrower 240
   see also migration
distress calls 182, 199
diving beetles 114
dragonflies 212
drink patch 59
drought 60, 225, 228, 233
Dryophytes 60, 76
Dryophytes chrysoscelis 60, 197, 212–13
Dryophytes cinereus 47, 64–5, 88
Dryophytes versicolor 212
Dusky Gopher Frog 229, 242–3
Duttaphrynus melanostictus 250, 268–9
dwarf clawed frogs 178, 179, 188–9

Eastern Gray Treefrog 212
ectothermy 62
Edalorhina perezi 198
effective population size 229
eggs 108–15
   chemical defenses 200
   clutch size 114
   on land 110, 112–13
Eleutherodactylus coqui 58
Eleutherodactylus jasperi 111
Eleutherodactylus johnstonei 152, 164–5
Eleutherodactylus martinicensis 164
Endangered Species Act 1973 260
environmental DNA (eDNA) 265
Epidalea calamita 92–3, 125, 145, 196, 233, 246–7
Espadarana prosoblepon 84, 100–1
estivation 60, 62, 70, 90
Euphlyctis hexadactylus 173
European Common Frog 32–3, 89, 123, 125, 148, 154, 199, 233, 255, 261
European Common Toad 30, 92, 104–5, 114, 124, 143, 148, 200, 233, 255
European Union 260
evolution 20–3, 26–8, 46
   arms race 193
   coevolution 153
   convergent 27, 46, 59, 171, 201
explosive breeding 89, 90–5, 96, 104
extinctions 11, 232, 250, 253, 264, 272

feeding 48, 166–89
   tadpoles 120–2, 124, 128
   see also predation by frogs
fighting 84–5, 100, 102
fire-bellied toads 199
fires 162
floating meadows 56–7
flowing water 38, 54–7, 83, 122–3, 132
flying frogs 140
foam nests 113, 117
foot adaptations 46–7, 53
founder effect 229
Fowler's Toad 225
freezing 60, 72
Fringed Leaf Frog 46
froglets 108, 111, 118, 196
   compass orientation 148
   dispersal 141

Gardiner's Seychelles Frog 168
gastromyzophory 132
Gastrotheca 113
genetic diversity 89, 229, 233, 242, 246, 264, 276
genetic rescue 233, 246
Genetically Modified Frogs (GMFs) 264
genomics 264
Geocrinia alba 266
Geocrinia vitellina 266
Geographical Information Systems (GIS) 262
giant frogs 31, 77
glassfrogs 84
gliding 47, 158
Golden Frog 111
Golden Mantella 257
Golden Poison Dart Frog 202
Goliath Frog 31
Green and Black Poison Frog 150
Green Frog 143, 160–1, 200
green frogs 154
Green Toad 145
growth inhibitors 125
growth rings 62, 224
Guinea Snout-burrower 171

habitat change 11, 36, 141, 144–5, 162, 242, 254, 272, 274
habitat fragmentation 230–1, 233, 254, 270
Hamptophryne boliviana 186
harlequin frogs 255
Harlequin Mantella 260, 270–1
headstarting 242, 274
hearing 81, 176
Hemisus 177
Hemisus guineensis 171
Hemisus marmoratus 225, 240–1
herbivores 172–3
Heterixalus 30
hibernation 60, 62, 90, 224
Horned Frog 123
horned frogs 175, 199
Huia cavitympanum 55
Hula Painted Frog 262
hybridization 270
hylid treefrogs 56, 60, 84, 255
Hymenochirus 179
Hymenochirus boettgeri 178, 188–9

Imitating Poison Frog 175, 209, 218–19

in vitro fertilization (IVF) 242
Indian Green Frog 173
individual identification 236
inertial suction 178–9
insectivores 168
introductions 149, 152–4, 164, 250
   see also reintroductions
invasive species 149, 254
iridophores 194
Itambé Bromeliad Frog 45, 143
IUCN 262, 272
Izecksohn's Brazilian Treefrog 173, 184–5

Johnstone's Whistling Frog *see* Lesser Antillean Frog
jumping 138–40

Kihansi Spray Toad 266, 272–3
Kouni Valley Striped Frog 170
Krefft's River Frog 82

Lacey Act 1900 260
Laguna de los Pozuelos' Rusted Frog 179
Larut Torrent Frog 132–3
lateral line organs 52, 53, 178
Latonia nigriventer 262
Laurel and Hardy effect 197
Leiopelma archeyi 36–7
leiopelmatid frogs 26, 28, 36–7, 76–7, 94, 140
leks 95, 134
Leopard Frog 253
Lepidobatrachus laevis 122, 182–3
leptodactylid frogs 81, 84, 89, 117
Leptodactylus fallax 117
Lesser Antillean Frog 152, 164–5
Limnonectes 84
Limnonectes palavanensis 134–5
Lines of Arrested Growth (LAGs) 224

INDEX

Linnaeus, Carolus 256
Lithobates catesbeianus 95, 96–7, 152, 200, 250
Lithobates clamitans 143, 160–1, 200
Lithobates pipiens 253
Lithobates sevosus 229, 242–3
Lithobates sylvaticus 60, 72–3, 144, 196, 238
locomotion 46–7, 136–65
Long-nosed Horned Frog 13, 205, 214–15
lungless frog 54

machine learning 236
Madagascar Bright-eyed Frog 80
magnetic field 148
Malagasy reed frogs 30
Mallorcan Midwife Toad 88, 102–3, 197, 231, 266
Mantella 201
Mantella aurantiaca 257
Mantella baroni 270
Mantella cowanii 260, 270–1
mantellid frogs 175, 201, 257, 260, 270–1
Marbled Snout-burrower 225, 240–1
Marsh Frog 148, 154
melanophores 194
metapopulation 230
Mexican Burrowing Toad 171, 180–1
Microhyla borneensis 45
microhylid frogs 14, 31, 45, 81, 94, 122, 175, 177
migration 141–50
    see also dispersal
mimicry 204, 209, 218
molecular genetics 229, 231
monkey treefrogs 47
Mountain Chicken Frog 117
Mountain Yellow-legged Frog 233, 244–5
mouth gaping 199
Müllerian mimicry 209, 218

Mutable Rainfrog 205, 216–17
myrmecophagy 171, 202

narrow-mouthed frogs
    see microhylid frogs
Natterjack Toad 92–3, 125, 145, 196, 233, 246–7
Nectophrynoides 111
Nectophrynoides asperginis 266, 272–3
Neobatrachia 15, 16, 28, 177, 179
Neobatrachus aquilonius 59, 70–1
Nicaragua Giant Glassfrog 84, 100–1
Nimbaphrynoides occidentalis 111
Northern Burrowing Frog 59, 70–1
Northern Corroboree Frog 276
Northern Cricket Frog 76, 148
Northern Gastric Brooding Frog 118

odor cues 148
Odorrana amamiensis 265
Odorrana splendida 265
Odorrana tormota 55
olfactory cues 176
Oophaga pumilio 117, 128–9, 202
Orange-bellied Frog 266
Oregon Spotted Frog 266
Oreophrynella nigra 199
Ornate Chorus Frog 224–5
Ornate Horned Frog 256–7
osteoderms 156, 214
Otton Frog 265
overexploitation 254
oviparity 111
ovoviviparity 111

Pac-man frogs *see* horned frogs
Paedophryne amauensis 31, 168
Panamanian Golden Frog 83, 98–9
Paracassina kounhiensis 170

parental care 114, 116–19, 130, 134, 218, 240
parotoid glands 34, 194
Passive Integrated Transponders (PIT) 236–7
pathogens 113, 190–219, 222, 250, 254–5, 266
    coevolution 153
    see also chytrid fungus
Pebble Toad 199
Pelobatrachus nasutus 13, 205, 214–15
Pelophylax 154
Pelophylax ridibundus 148, 154
Perez's Snouted Frog 198
pest control 149
pet trade 152, 182, 188, 202, 254, 255, 260, 270
Phrynobatrachus krefftii 82
Phyllobates aurotaenia 202
Phyllobates bicolor 202
Phyllobates terribilis 202
Phyllomedusa 47
Phyllomedusa bahiana 198
Physalaemus pustulosus 176
Pipa pipa 113, 126–7, 178
pipid frogs 52–3, 77, 113, 126–7, 178, 179
pitcher plants 45, 114
poison dart frogs 128, 202
poison frogs 141, 150, 175, 201–2, 208, 255
pollinators 184
population bottlenecks 229
population estimates 234–7
predation by frogs 53, 168
    ambush 174–5
    inertial suction 178–9
    see also feeding
predators of frogs 52, 123–4
    defenses against 190–219
    tags 237
Pristimantis mutabilis 205, 216–17
prolonged breeding 90–5
Pseudacris 60
Pseudacris crucifer 45, 60, 212

Pseudacris maculata 224, 238–9
Pseudacris ornata 224–5
Pseudhymenochirus 178
Pseudophryne 202
Pseudophryne corroboree 266, 276–7
Pseudophryne pengilleyi 276
public engagement 256–61
pumpkin toadlets 140, 156–7
purple burrowing frogs 28, 40
Pyxicephalus adspersus 84, 119, 130–1

rain frogs 111
Rana dalmatina 237, 274–5
Rana luteiventris 231
Rana muscosa 233, 244–5
Rana pretiosa 266
Rana sierrae 244
Rana temporaria 32–3, 89, 123, 125, 148, 154, 199, 233, 255, 261
ranaviruses 255
Ranitomeya fantastica 218
Ranitomeya imitator 175, 209, 218–19
Ranitomeya summersi 218
Ranitomeya variabilis 218
Red List 262
Red Pumpkin Toadlet 140, 156–7
Red-eyed Treefrog 197, 208, 210–11
reintroductions 242, 244, 246, 266, 272, 274, 276
reproduction 48, 106–35
    "boom and bust" 224, 240
    communication 76–83, 87–8
    explosive breeding 89, 90–5, 96, 104
    multiple paternity 89
    new ponds 144–5
    prolonged breeding 90–5
    sexual selection 86–7
    see also tadpoles

Rhacophorus nigropalmatus 140, 158–9
Rheobatrachus silus 118
Rheobatrachus vitellinus 118
Rhinella diptycha 170, 173
Rhinella marina 149, 152, 170, 176, 250
Rhinoderma darwinii 117
Rhinoderma rufum 117
Rhinophrynus dorsalis 171, 180–1
River-swamp Frog 170
Rock Frog 197
Rocky Mountain Tailed Frog 38–9
role reversal 88, 102, 134

Sabah Huia Frog 55
Sabah Splash Frog 82
salt tolerance 64
Sanyan Frog 23
satellite males 88, 96
Scaphiopus couchii 59, 68–9
Sechellophryne gardineri 168
seed dispersal 184
semicircular canals 156
sexual selection 86–7
Seychelles Frog 6, 28, 40–1
Seychelles Treefrog 30
shrinking 198
Sierra Nevada Yellow-legged Frog 244
sink population 230
size 171
　inflation 13, 182, 198
　largest and smallest frogs 31
skeletochronology 224
skin 156, 194
　defenses 192–4, 203
　drink patch 59
Surinam Toad 126
　see also coloration; toxins
skulls 171
slug specialists 170
Smith, Edward Percy 154
Smooth Guardian Frog 134–5
sneaky mating 88, 89

snouted-burrowers 177
sooglossids 6, 28, 40–1, 81
Sooglossus sechellensis 6, 28, 40–1
source population 230
Southern Corroboree Frog 266, 276–7
Southern Cricket Frog 148
Southern Gray Treefrog 197, 212–13
Southern Toad 170, 176
splash frogs 77, 82
Splendid Treefrog 46
Spring Peeper Frog 45, 212
startle response 198, 208
Staurois 77
Staurois guttatus 82
Staurois latopalmatus 82
Strawberry Poison Frog 117, 128–9, 202
Summers' Poison Frog 218
Surinam Horned Frog 175, 186–7
Surinam Toad 113, 126–7, 178
Surinam toads 52, 113, 126–7, 178
swimming 140
Sylvia's Treefrog 46

Tachycnemis seychellensis 30
tadpoles 58, 108, 111, 120–5, 205, 222
　anti-predator behaviors 196–7
　chemical defenses 200
　development 120–2, 196–7
　feeding with infertile eggs 117, 128, 218
　flowing water 54
　gastromyzophory 132
　parental care 116–19
　in plants 44, 45
　Rocky Mountain Tailed Frog 38
　semiterrestrial 197
　temporary ponds 50–1, 68, 119, 124

tags 235, 236–7
　VIE tags 237
telmatobiid frogs 53, 66–7, 179
Telmatobius culeus 53, 66–7
Telmatobius macrostomus 53
Telmatobius rubigo 179
Temnospondyli 21–2
temperature
　extreme 58–62, 64
　tadpoles 123
temporary ponds 48–51, 68, 119, 124
Thoropa taophora 197
Three-striped Poison Frog 150
Tiger Salamanders 238
Tiktaalik roseae 21
Titicaca Water Frog 53, 66–7
"Toads on Roads" campaigns 141, 257
toe pads 46–7, 132, 158
tongues
　absence 126, 178
　protraction 177–8
torrent frogs 123
toxins 13, 34, 52, 200–3
　coloration warning 98, 156, 175, 193, 198–9, 200, 204–5, 208, 218
tracking frogs 149, 150, 160, 231, 242
treefrogs 46–7, 76, 95, 158
　American Green Treefrog 47, 64–5, 88
　Asian 47
　Bahia Green Treefrog 198
　Eastern Gray Treefrog 212
　hylid 56, 60, 84, 255
　Izecksohn's Brazilian Treefrog 173, 184–5
　monkey treefrogs 47
　Red-eyed Treefrog 197, 208, 210–11
　Seychelles Treefrog 30
　Southern Gray Treefrog 197, 212–13
　Splendid Treefrog 46

Sylvia's Treefrog 46
Triadobatrachus massinoti 22–3
tubercles 216
Túngara Frog 176

ultrasound calls 55
unken reflex 198–9

Variable Poison Frog 218
Visible Implanted Elastomers (VIE tags) 237
visual communication 82–3, 87, 98
viviparity 111
vocal sac 77, 82, 117
vocalization 76–81, 83, 87–8, 96, 104, 156
　breeding chorus 93, 95
　distress calls 182, 199
　Seychelles Frog 40
　water noise 55

Wallace's Flying Frog 140, 158–9
water bodies 48–57
Water-holding Frog 60
weapons 84–5, 100
webbed feet 53
Western Nimba Toad 111
Western Toad 145, 162–3
White-bellied Frog 266
Wood Frog 60, 72–3, 144, 196, 238
World Congress of Herpetology 253

xanthophores 194
Xenohyla truncata 173, 184–5
Xenopus laevis 13, 122, 124, 178

"Year of the Frog" 257

ACKNOWLEDGMENTS

We thank Richard Webb, Wayne Blades, Slav Todorov, Kate Shanahan, Elaine Willis, John Woodcock, Les Hunt, and all the team at UniPress Books for diligently working with and supporting us throughout the process of putting this fantastic book together. We also offer our extended thanks to the following people for their valued input into Lives of Frogs: Andrius Pašukonis, Ben Tapley, Brett Lewis, Chien Lee, Christoph Liedtke, Chun G. Kamei, David Fielding, Devin Edmonds, Franco Andreone, Gerardo Garcia, Gonçalo Rosa, Izabela M. Barata, Jason Isaacs (hello), Jeffrey W. Streicher, Joseph Trafford, Kay Bradfield, Les Minter, Richard Gibson, Stephen Mahony, Thomas Doherty-Bone.

ILLUSTRATION AND PICTURE CREDITS

P16: Hime P. M. et al. (2021). https://doi.org/10.1093/sysbio/syaa034 (Fig. 1); P22: Pough, F. H., et al. (2013). Vertebrate Life (9th ed.). (Pearson) (Fig. 10-5); Gao, K-Q, and Y. Wang. (2001). https://doi.org/10.1671/0272-4634(2001)021[0460:MAFLPC]2.0.CO;2 (Fig. 5); P23: Ascarrunz, E. et al. (2016). https://doi.org/10.1163/18759866-08502004 (Fig. 1); P34: Wilber, C. G., and P. L. Carroll. (1940). https://doi.org/10.2307/3222820 (Plate 1 [2]); Jared, C., et al. (2009). https://doi.org/10.1016/j.toxicon.2009.03.029 (Fig. 1 [4]); P47: Channing, A., et al. (2016). https://doi.org/10.11646/zootaxa.4155.1.1 (Fig. 18); Cruz, C. A. G., and U. Caramaschi. (2003) https://www.researchgate.net/publication/306196483_Taxonomic_status_of_Melanophryniscus_stelzneri_dorsalis_Mertens_1933_and_Melanophryniscus_stelzneri_fulvoguttatus_Mertens_1937_Amphibia_Anura_Bufonidae (Fig. 5); de Medeiros Magalhães, F., et al. (2014). https://doi.org/10.1655/herpetologica-d-13-00054 (Fig. 2 D); Duellman, W. E. (1970). https://www.biodiversitylibrary.org/bibliography/2835 (45 B, 167 E); Formas, J. R., et al. (2006). https://doi.org/10.1655/05-08.1 (Fig. 2 C); Goldberg, J., and M. Fabrezi. (2008). https://doi.org/10.1111/j.1096-3642.2007.00345.x (Fig. 2 L); Illinois Natural History Survey Herpetology Collection. (2024). https://herpetology.inhs.illinois.edu/species-lists/key-to-Illinois-species/key-to-frogs-and-toads-of-illinois (Fig. 9); P53: Claas, B., and H. Münz. (1996). https://doi.org/10.1007/BF00188167 (Fig. 2); Shelton, P. M. J. (1970). https://doi.org/10.1242/dev.24.3.511 (Fig. 2); P56: Upton, K., et al. (2014). https://ssarherps.org/herpetological-review-pdfs/ (Fig. 1); P77: Duellman W. E., and Trueb L. (1994) Biology of Amphibians (McGraw-Hill) (Fig. 4.9, 4.10); Gans, C. (1973). https://doi.org/10.1093/icb/13.4.1179 (Fig. 1, 2, 4, 6); P80: Labisko, J. (unpublished); P81: Hetherington, T. E. (1992). https://doi.org/10.1007/978-1-4612-2784-7_25 (Fig. 21.2); Duellman W. E., and Trueb L. (1994) Biology of Amphibians (McGraw-Hill) (Fig. 4.11); P94: Vargas-Salinas, F., et al. (2020). https://doi.org/10.1093/biolinnean/blaa009 (Fig. 3.7); P111: Wells, K. D. (2007) The Ecology and Behavior of Amphibians (The University of Chicago Press).

(Fig. 10.4 A, B, E; 10.5 C; 10.7 B; 10.8 A); P121: Duellman W. E., and Trueb L. (1994) Biology of Amphibians (McGraw-Hill) (Fig. 6.12 B, C, F; 6.14 A); Duellman, W. E. (1970). https://www.biodiversitylibrary.org/bibliography/2835 (Fig. 214); Liu, Chengzhao. (1950). https://www.biodiversitylibrary.org/bibliography/2977 (Fig. 38 A); Tapley, B. et al. (2020). https://doi.org/10.11646/zootaxa.4845.1.3 (Fig. 4A); Can't find reference for 'Ceratophrys aurita carnivorous'; P124: Löschenkohl, A. (1986) Proceedings of the 3rd Annual General Meeting of Societas European Herpetologica (Fig. 1); P132: Matsui, M., and J. Nabhitabhata. (2006). https://doi.org/10.2108/zsj.23.727 (Fig. 2C); P140: Pough, F. H., et al. (2013). Vertebrate Life (9th ed.). (Pearson) (Fig. 10-6); P141: Sinsch, U. (1990). https://doi.org/10.1080/08927014.1990.9525494 (Fig. 3); P149: Slezak, M. (2014) https://www.newscientist.com/article/mg22229660-600-aliens-versus-predators-the-toxic-toad-invasion; Urban, M. C., et al. (2008). https://doi.org/10.1086/527494 9 (Fig. 1); P158: McCay, M. G. (2003). https://doi.org/10.1646/0006306(2003)035[0094:WUTRFC]2.0.CO;2 (Fig. 1B); P170: Drewes, R. C., and Roth, B. (1981). https://doi.org/10.1111/j.1096-3642.1981.tb01573.x (Fig. 2, Fig. 3); P171: Paluh, D. J., et al. (2020). https://doi.org/10.1073/pnas.2000872117 (Fig. 4 [13, 14, 15, 25, 26, 29]); P177: Gans, C., and G. C. Gorniak. (1982). https://doi.org/10.1126/science.216.4552.1335 (Fig. 1, 2); Pough, et al. (2016). Herpetology (4th ed.). Sinauer Associates, Incorporated. (Fig. 11.12); P179: Carreno, C. A., and K. C. Nishikawa. (2010). https://doi.org/10.1242/jeb.043380 (Fig. 1); Sokol, O. M. (1969) http://www.jstor.org/stable/3890988 (Fig. 7A, B; Fig. 8A, B, C, D); P188: Deban, S. M., and W. M. Olson. (2002). https://doi.org/10.1038/420041a (Fig. 1); P197: Griffiths, R. (unpublished); P198: Toledo, L. F., et al. (2011). https://doi.org/10.1080/03949370.2010.534321 (Fig. 2); P216: Guayasamin, J. M., et al. (2015). https://doi.org/10.1111/zoj.12222 (Fig. 2); P224: Sinsch, U., et al., (1997). https://www.thebhs.org/publications/the-herpetological-journal/volume-17-number-2-april-2007/486-10-growth-marks-in-natterjack-toad-i-bufo-calamita-i-bones-histological-correlates-of-hibernation-and-aestivation-periods; P229: Trafford, J. (unpublished); P252: Luedtke, J., et al. (2023). https://doi.org/10.1038/s41586-023-06578-4 (Fig. 2); P263: Licata, F., et al. (2023). https://doi.org/10.1038/s41598-023-29467-2 (Fig. 2).

Alamy Photo Library: Ken Griffiths 14R; Daniel Heuclin / Biosphoto 15; Nature PL / MYN / JP Lawrence column 1 17T; Nature PL / Chris Mattison column 1 17M; Stuart Wilson / Biosphoto column 1 17B; Chris Mattison column 2M; Nature PL / MYN / JP Lawrence column 2 17B; Nature PL / MYN / Clay Bolt column 3 17B; Nature PL / Edwin Giesbers column 4 17B; Daniel Heuclin / Biosphoto 31B; All Canada Photos 39; Michael Doolittle 44; Arterra Picture Library 50; Jan Carroll 58; Robert Hamilton 59; Skip Moody 61T; FLPA 62; Chris Mattison 63B; Bert Willaert 67; Ondrej Prosicky 80T; Matthijs Kuijpers 82; John Cancalosi 86; Premaphotos 87; Chris Mattison 103; Michael & Patricia Fogden/ Minden 112R; Daniel Heuclin / Biosphoto 127; Ernst Dirksen Minden Picture 139; Geoff Du Feu 144; Frédéric Grimaître 145; Fero Bednar 146; Sean Cameron 148T; John Cancalosi 153; Sam Yue 159; Paul R. Sterry 165; Jelger Herder 168; Daniel Heuclin 169; Daniel Heuclin 174; Adrian Hepworth 181; Nick Garbutt 215; Morley Read 217; Anton Sorokin 219; Mark Payne-Gill 232; Cyril Ruoso 237; PureStock 245; Lucas Bustamante 250-251; Paralaxis 252; Francisco Martinez 261; Cyril Ruoso 266-267; Ken Griffiths 277;
Robin James Backhouse: 241
João Burini: 157
Chien C. Lee: 7; 8; 9; 10; 14L; 45; 55; 77; 92R; 95; 110R; 113; 116;117; 119; 129; 135; 142; 170; 187; 195; 201T; 206; 208T; 208B; 209R; 209L; 227;
Dreamstime.com Tatiana Saenko: 183
Devin Edmonds: 271
Vincent Egan / Les Minter: 84-85
Richard Gibson Auckland Zoo: 36
Richard Griffiths: 256
Getty Images: Auscape 71; freder 99.
Jim Labisko: 6; 28
Brett Lewis Photography (Instagram - @lensviper): 29L; 29R; 41; 247
Nature Picture Library: Lucas Bustamante 46; Andy Sands 47; D. Parer & E. Parer-Cook 63T; Rolf Nussbaumer 69; Chris Mattison 80B; Charlie Hamilton-James 88; Eric Medard 92L; Shane Gross 123; Cyril Ruoso 141; Kim Taylor 175; Chris Mattison 178; Chris Mattison 192; Christian Ziegler 196; Barry Mansell 197; Bert Willaert 198; Chris Mattison 204; Piotr Naskrecki 225; Solvin Zankl 231R; Oriol Alamany 231L; Pete Oxford 233R; Nick Garbutt 260; Melvin Grey 275
Henrique Nogueira: 173; 185
Andrius Pašukonis: 150; 151
Uzi Paz / Pikiwiki Israel: 262
Delfi Sanuy: 224
Jake Scott: 243
Benjamin Tapley / ZSL: 108; 228;
Shutterstock: OleksiGS 2; OleksiGS 3; OleksiGS 4TL; Kurit afshen 4TR; Eric Isselee 4M; Richard Peterson 4B; Rosa Jay 5TL; Eric Isselee 5TR; Eric Isselee 5BL; Nynke van Holten 5BR; Dirk Ercken 9; Eric Isselee 10; Michiel de Wit 13; Chase D'animulls column 2 17T; taviphoto column 3 17T; Nynke van Holten column 3 17M; Michael de Wit column 4 17T; natthawut ngoensanthia column 4 17M; Danny Ye 21; Oleg Kovtun Hydrobio 24; Dani Jara 25; Matteo photos 27; zdenek macat 33; Amy Lutz 35; Heying Hua 49; Nick Greaves 51; Diego Paramo 52; Murrrrr-s 54; Ken Griffiths 60; Elementspace 61B; Lorraine Hudgins 65; PhotoZoomstock 73; Sushil Prajapati 76; worldswildlifewonders 78; Kristine Rad 79; Steve Bower 81; Yashpal Rathore 83; Monika Surzin 89; Marco Maggesi 91; Peter K. Ziminski 93; Ilias Strachinis 97; Ana Dracaena 101; Miroslav Hlavko 105; Matteo photos 109; Rosa Jay 110L; Reality Images 112L; Victor Mateo 114-115; Stephanie Periquet 118; Kurit Afshen 120-121; Rosa Jay 122; Ian Redding 125; Milan Zygmunt 131; Trygve Finkelsen 148B; Jay Ondreicka 152; Alexander Denisenko 154-155; Tom Reichner 161; Wirestock Creators 163; Nilanka Sampath 172; Vladimir Turkenich 176; Dan Olsen 189; Agus Gatam 193; elakazal 200; Julen Arabaolaza 201B; Oleksandr Kostiuchenko 203; Michiel de Wit 205; Dirk Erckin 207; K Hanley CHD Photo 209R; Vaclav Sebek 211; Jeff Weisbecker 213; Petr Muckstein 222-223; Birdtolk 226; Coulanges 233T; SciPhi.tv 234; Zoltan Tarlacz 235; Leonardo Mercon 236; Paul Reeves Photography 239; Anupong Termin 257B; Lefteris Papaulakis 257T; Jkuy 259; Jubal Harshaw 264; Yuya Wakita 265; Pakin Techaphiarrat 269; David W. Leindecker 273; Kurit afshen 278-279
Wiki Commons: Eduardo Ascarrunz, Jean-Claude Rage, Pierre Legreneur, Michel Laurin CC BY 3.0 23; Stephan Sprinz CC-BY-4.0 26; Renato Augusto Martins CC BY-SA 4.0 202